Mineral Processing Technology

Mineral processing technology is a branch of applied science that deals with the principles and practice of separating useful minerals from primary solid ore mineral resources. This book introduces the science and technology of processing solid minerals to concentrates of grades suitable for industrial extraction of metal values and other non-metallic products. It also includes case studies, typical process flowsheets, aspects of the processing of tailings arising from mineral processing plants and worked examples.

Features:

- Includes science and technology of processing solid minerals to concentrates of grades, suitable for industrial extraction of metal values and other non-metallic products.
- Provides a logical progression from basic to advanced concepts in mineral processing.
- Designed to stimulate students to think as mineral processing engineers in training.
- Explores sustainable mineral processing and circular economy in mineral processing.
- Contains worked examples that clearly illustrate the various theories presented and help readers develop problem-solving skills in mineral processing.

This book is aimed at professionals and senior undergraduate students in metallurgy, mining, mineral processing, chemistry and chemical engineering.

Mineral Processing Technology
A Concise Introduction

Abraham Adewale Adeleke

CRC Press
Taylor & Francis Group
Boca Raton London New York

CRC Press is an imprint of the
Taylor & Francis Group, an **informa** business

Designed cover image: © Shutterstock

First edition published 2023
by CRC Press
6000 Broken Sound Parkway NW, Suite 300, Boca Raton, FL 33487-2742

and by CRC Press
4 Park Square, Milton Park, Abingdon, Oxon, OX14 4RN

CRC Press is an imprint of Taylor & Francis Group, LLC

© 2023 Abraham Adewale Adeleke

ISBN: 978-1-032-34704-2 (hbk)
ISBN: 978-1-032-34705-9 (pbk)
ISBN: 978-1-003-32343-3 (ebk)

DOI: 10.1201/9781003323433

Typeset in Times LT Std
by Apex CoVantage, LLC

This book is dedicated to the glory of God, the Almighty, the Giver of Life and the Source of wisdom and knowledge.

Contents

About the Author

Abraham Adewale Adeleke is a Professor and a former head of the Department of Materials Science and Engineering, Obafemi Awolowo University, Ile Ife, Nigeria. He has over 30 years' experience in metallurgical research and University education. He was a research metallurgist for about 15 years during which he rose to the position of Chief Metallurgical Engineer and he has been a Professor since 2015. He has taught courses such as mineral processing and hydrometallurgy for several years in both Nigeria and South Africa. During his temporary residence in South Africa, he was a research fellow at the Tshwane University of Technology, Pretoria and an Associate Professor at the Vaal University of Technology, Vanderbijlpark.

Furthermore, Abraham has over 40 research articles and contributions of chapters in books to his credit. He has also contributed three articles as a co-author in two proceedings of the US-based Mineral, Metals and Materials Society (TMS) Technical Meetings.

Professor Adeleke has served as an External Examiner for postgraduate programmes at several universities in Nigeria and South Africa. He was an in-plant trainee in coal research under the United Industrial Development Organization (UNIDO) Fellowship in Romania in 1994. He is a registered Engineer with the Council for the Regulation of Engineering in Nigeria (COREN) and he was registered with the Southern Africa Institute of Mining and Metallurgy (SAIMM) and South Africa Coal Processing Society (SACPS) during his stay in South Africa.

Preface

Mineral processing refers to the science and technology of processing a solid mineral's run-of-mine (ROM) using a combination of unit operations to obtain an upgraded form called concentrate which is suitable for further industrial processing such as for smelting to produce metals and their alloys. The technology of mineral processing is the third step in the relay race which commences with geological exploration and proceeds to mining, mineral processing, extractive metallurgy and terminates in production metallurgy/engineering to produce finished metallic and non-metallic products.

This book is a concise introduction to mineral processing technology. It is a product of more than thirty years of research and teaching in the subject areas of mineral/coal processing and hydrometallurgy. It is an attempt to build on the foundation laid by pioneers and renowned academics like Dr Barry A. Wills and others. The book is unique for several reasons. Firstly, it emphasizes the place of geology and geological explorations as the starting point of mineral processing and extractive metallurgy. Secondly, it is structured based on the author's experience to make it appealing to first-time students of the course who had no previous knowledge of solid minerals and their processing.

Thirdly, there are illustrative numerical examples and diagrams included, where applicable, to further assist the students to grasp the basic concepts and theories of the course. Fourthly, the book introduces new concepts in mineral processing such as geometallurgy and circular economy in a way first-time students of the course can comprehend them. The fifth unique aspect of the book has to do with the presentation of summaries of selected bench- and pilot-scale research studies that involved the author and some other workers to further improve readers' understanding of the subject.

Finally, the content of the book covers virtually all topics in mineral processing. The book is recommended for undergraduate and postgraduate students of science, engineering and technology particularly those in Mineral Processing Engineering, Mining Engineering, Materials Science, Materials Engineering, Metallurgy, Chemical Engineering and Chemistry.

1 General Introduction

1.1 Introduction

A mineral is a naturally occurring crystalline, inorganic solid found in the earth's crust and some other locations on the earth and even beyond the planet earth in other extra-terrestrial bodies such as asteroids and having definite chemical compositions. Minerals are crystalline because of their long-range atomic order. Minerals can be categorized broadly into four: industrial minerals, mineral fuels, rock and ores. Industrial minerals refer to minerals such as diamond, corundum, phosphate and dolomite; mineral fuels refer to coal and oil sands; rocks are sources of aggregate, sand and gravel while ores are mineral deposits with grades and reserves considered adequate to qualify them as sources from which metals can be economically extracted.

For instance, galena, the metallic ore mineral that bears the metal lead usually occurs as an aggregate or a mixture of lead, zinc, iron and silicon compounds with chemical formulae PbS, ZnS, FeS_2, SiO_2 and Fe_2O_3. Mineral raw materials are the primary sources from which metals of industrial importance such as iron, aluminium, copper and zinc as well as non-metals like glass and refractory bricks are produced. Mineral materials such as coal, petroleum and radioactive elements like uranium are also important energy sources.

Mineraloids are naturally occurring solid non-crystalline or amorphous materials found in the mineral kingdom but possess chemical compositions that vary beyond ranges generally accepted for specific minerals or they are non-crystalline in structure. They can also be described as naturally occurring materials that unlike crystalline materials lack long-range atomic order. They may be categorized into two groups—the organic group containing carbon and the inorganic group with no carbon content. Coal is an example of a mineraloid of the organic type, while opal is an inorganic type.[1]

Coal is a non-crystalline, organic substance without a fixed chemical composition due to its biogenic origin. In view of this, though it is in the mineral kingdom, it is strictly not a solid mineral but a mineraloid. Similarly, the chemical formulae of opal and obsidian are variable and not definite as true minerals. Obsidian is a mixture of 70–75% SiO_2 plus MgO and Fe_3O_4, while opal has the formula $SiO_2.nH_2O$ showing their indefinite chemical compositions. Synthetic minerals are minerals that are produced in the laboratory through chemical processes and not naturally occurring like true minerals. Artificial minerals refer to materials such as slags, concretes and mill scales that result from the processing of natural minerals or their metallic products.

Earth's crust is the outermost surface layer of the planet. The earth's crust is made up of diverse types of sedimentary, igneous and metamorphic rocks. The mantle lies under the earth's crust and its upper part is made up mainly of a rock called peridotite,

DOI: 10.1201/9781003323433-1

with a density greater than rocks commonly found in the overlying crust. The oceanic crust is about 4–6 km thick, while the continental crust is about 32 km in thickness.

Minerals have been part of man's culture from the use of flints during the Stone Age to the use of uranium ore during the atomic age. The origin of ore minerals can be traced to the elements found originally in nature as cations and anions, such as Na^+, Ca^{2+}, O^{2-}, Si^{4+}, Fe^{2+} and CO^{2-} which under heat and pressure transform to simple inorganic minerals like SiO_2, Fe_2O_3 and $CaCO_3$. Under further heat and pressure, the simple minerals combine to form complex minerals call rocks. The rocks under heat, pressure, deformation and chemical activities got ores from which metals or other non-metallic products can be derived formed in them.[2]

Mineral processing can be defined as a sequence of unit operations coordinated to treat solid mineral resources in order to reduce the unwanted gangue minerals they contain and thus relatively increase the mineral value(s) of interest and thereby produce two products, a concentrate that consists mainly of the mineral value(s) and a tailing comprising mostly of the gangue minerals. The equipment used in mineral processing carry out unit operations, that is, they each perform a single task only and the material is passed to the next equipment in sequence.

1.2   Locations of Minerals

There are two categories of mineral deposit locations. These are:

1. Terrestrial Locations
   Solid minerals are found on the planet earth in the following locations:
   a. Earth's crust-on the surface, sub-surface and underground.
   b. Seawater—Mg, Li and salts are found dissolved in seawater.
   c. Sea bed—some solid minerals are found in sea bed ridges. The sea floor contains ores that host high-density metals such as titanium, tin and gold and other heavy non-metallic materials like diamond. These are mechanically eroded from exposed rocks on land and become concentrated in pockets by flowing water on the sea floor. In addition, phosphorite and manganese minerals are chemically precipitated in the form of nodules and crusts from materials dissolved in seawater.
   d. Ocean beaches—beaches also host solid minerals such as rutile and quartz.
2. Extra-terrestrial Locations

Bodies outside the planet earth such as asteroids also host precious metals such as gold, platinum group metals (PGMs), nickel, cobalt, aluminium and titanium.

1.3   Mineral Deposits

A mineral resource is any material of value that can be extracted from the earth's crust. The mineral resources can be categorized as follows:

1. Gaseous resources
   These are resources such as natural gas

2. Liquid resources
   These are resources such as crude oil
3. Semi-solid resources
   These include the semi-solid tar sand
4. Solid resources

These include metallic and non-metallic minerals. An ore is an aggregate of minerals with one or more predominating and thus determining the name of the mineral. A mineral deposit is a natural accumulation of accessible useful solid minerals formed due to geological actions such as magma eruption, flow of rainwater, motion of wind and groundwater movement. On the other hand, an ore deposit is a mineral deposit that has an acceptable reserve size, acceptable grade, acceptable mine-ability and location depth to allow for profitable recovery of its value mineral(s). For instance, the clay mineral contains aluminium but it is not considered an ore mineral for aluminium extraction because its content of aluminium is low and the nature of its occurrence as complex silicates in the clay is such that economic extraction is not feasible with the present technology. On the other hand, bauxite is considered an ore of aluminium because it contains aluminium as simple inorganic aluminium oxide and its aluminium content can be economically extracted from its associated gangue minerals by Bayer process and electrolysis in a molten state. Metalliferous ores refer to metallic ores containing or yielding metals and they include ores such as those of iron, copper, nickel, gold, zinc and uranium. The geochemical interest in ore deposits derives from their relative rareness and the relatively large contents of metals in them.[3-5]

An ore deposit is a natural accumulation of ore minerals in the earth's crust usually in a rock formation of large size of varying shapes. The quantity of the accumulated minerals must be sufficiently large in terms of their thickness and length in metres. For instance, a patch of mineral of about 0.5 m thick and 1.5 km long cannot constitute an ore deposit since the quantity of its mineral cannot pay for the cost of the extraction. The grade of the mineral is also an important factor to be considered in determining its definition as an ore deposit. For example, an iron ore mineral deposit of say 18% Fe cannot be considered an ore deposit for iron extraction since iron ores are high-grade low-value minerals with a typical run-of-mine grade as high as over 60% in some cases. The depletion of higher-grade iron ore deposits in the future will however render lower-grade iron deposits to become ore deposits suitable for economic use.

Ore minerals bearing a value mineral of interest are usually found in association with some other unwanted minerals, called gangue; particularly quartz which often constitutes the main undesirable mineral and some metallic minerals, which occur in smaller quantities that render them unsuitable to be called primary ores for their contained metals.

1.4　Categories of Solid Minerals

Solid minerals can be categorized into the following types:

1. Ferrous minerals
   These are minerals from which iron can be extracted. These include hematite (Fe_2O_3), magnetite (Fe_3O_4), pyrite (FeS_2), limonite ($FeO.nH_2O$) and

goethite (FeO(OH)). Ferrous alloys in the form of steel and its alloys are the most widely used structural material. In 2014, the world's output of steel was about 1.6 billion tons with China's production share being about 830 million tons. Steel alloys are widely used because of their good mechanical strength in terms of tensile and yield strength, good malleability and fair machinability.

2. Non-ferrous minerals

These are minerals from which metals other than iron can be extracted. They are classified as follows:

a. Ferroalloys minerals.

These minerals such as chromite, pyrolusite and quartz are usually smelted with iron ore or iron-based scraps to produce ferroalloys containing in this case the elements chromium, manganese and silicon to produce ferrochromium, ferromanganese and ferrosilicon, respectively. Other ferroalloy minerals include wolframite (for ferrotungsten), tantalite (for ferrotantalum) and columbite (for ferroniobium).

b. Base metal minerals are minerals bearing metals that are commonly used such as lead, zinc and copper with galena, sphalerite and cuprite as typical ores; respectively.

c. Light metal minerals.

Ore minerals such as bauxite, rutile and magnesite from which light metals such as aluminium, titanium and magnesium, respectively are extracted.

d. Metal fuel minerals are minerals that bear metals for energy generation. These include uranium and polonium. Examples of uranium minerals are pitch blende, uranitite and carnotite.

e. Noble metals are metals that do not easily corrode in dry or moist air, water and other corrosive environments. These include gold, silver and platinum groups of metals (PGMs). The PGMs refer to the metals platinum, palladium, rhodium, ruthenium, osmium and iridium.

3. Non-metallic minerals

Non-metallic minerals are minerals from which metals cannot be extracted economically based on the available technology. They are used in bulk the way they are found. They are classified as follows:

a. Construction minerals—minerals used in the construction of buildings and roadways. Examples are marbles, limestones and granites.

b. Refractory minerals—minerals that are used to produce refractory (heat-resistant) bricks for furnaces. Examples are magnesite, dolomite, fireclay, low-grade chromite and graphite.

c. Industrial minerals—minerals used for industrial purposes. Examples include barite for drilling mud, diamond for use as cutting tools, phosphate for fertilizer production, China clay as filler in paper making and gypsum for pharmaceutical applications.

d. Gemstones—categorized as precious and semi-precious. Examples of precious gemstones are emerald, ruby, diamond and sapphire while examples of semi-precious gemstones are agate, alexandrite, agate, amethyst, aquamarine, garnet, moonstone, opal, pearl, spinel, tanzanite, tourmaline, turquoise, chrysoprase and zircon.

Industrial minerals are used in various areas of applications in industries such as in making abrasives (quartz, corundum, diamond), filler and pigment (barite, calcite, bentonite, dolomite, feldspar), plastic (calcite, kaolin, talc, wollastonite, mica), ceramics (quartz, kaolin, feldspar) and fertilizers (phosphate, potash, calcite, dolomite).

1.5 Chemical States of Mineral Occurrence

The chemical states in which minerals are found in nature depend on the reactivity of their primary metal contents to oxygen, sulphur and carbon dioxide in the earth's crust. In view of this, minerals are found as follows:

1.5.1 Oxides

Metals that have a higher affinity for oxygen preferably react with oxygen and will be found in the earth's crust in oxide-stable form. Examples are hematite (Fe_2O_3), magnetite (Fe_3O_4), quartz (SiO_2) and cassiterite (SnO_2).

1.5.2 Sulphides

Some metals readily react with sulphur and will exist in the sulphide-stable form in the earth crust. Examples are galena (PbS), sphalerite (ZnS) and pentlandite (NiS).

1.5.3 Carbonates

Some metals readily react with carbon dioxide and are found in the carbonate stable form in the earth crust. Examples are siderite ($FeCO_3$), malachite ((Cu_2CO_3)(OH)$_2$), smithsonite ($ZnCO_3$) and cerussite ($PbCO_3$).

1.5.4 Arsenides

When a metal readily reacts with arsenic other than oxygen, sulphur and carbon dioxide, it forms an arsenide. Examples are platinum arsenide (PtAs2), niccolite (NiAs), skutterudite ($CoAs_3$) and maucherite ($Ni_{11}As_8$).

1.5.5 Native Noble Metals

Some metals do not readily react with sulphur, oxygen and carbon dioxide and occur in the earth crust as native metals. Native noble metals are gold, silver, copper and platinum group metals (PGMs). However, copper does not resist oxidation like other noble metals.

1.6   The Purpose of Mineral Processing

There are three main reasons why mineral processing is necessary:

1. Mineral processing is required to reduce the gangue minerals in an ore to be transported from the source mine to the smelting plant for metal recovery. The reduction in the weight of the ore for transport will reduce the cost of transport.
2. Mineral processing is also required to produce concentrates or in some cases super-concentrates for industrial use. Industrial metallurgical plants such as the blast furnace have specifications for the raw material inputs. For the blast furnace, the specification for the iron ore concentrate total Fe is 63% (minimum) and for the direct reduction Midrex process, the iron ore concentrate total Fe must be 66.5% (minimum) for efficient reduction and smelting to iron.
3. Furthermore, mineral processing allows the splitting of a complex ore with more than one mineral value into its component parts as concentrates and for separate collections. Thus, a complex sulphide ore of galena containing sphalerite and pyrite can be split into its components by froth flotation treatment. The separate value commodities can then be marketed differently for maximum returns.

1.7   Geological Exploration

Geological explorations are carried out at three broad levels, that is, regional, district and deposit scale. The task at the regional level includes a survey using aircraft for geophysical and geochemical studies, ground checking, collection of soil and rock chips, digging of pits and trenches, establishing of mineralization by few scout drillings, marking out of priority areas and ranking of selected targets. Mineralization refers to the process of deposition of metal values during the formation of ore bodies or lodes. In a geochemical survey, reconnaissance and detailed surveys are carried out. A reconnaissance survey requires sampling of stream sediments. Detailed surveys are then carried out using samples obtained from closely spaced points a few metres apart over an area spanning a few square kilometres. The purpose of the detailed survey is to identify and delineate a geochemical anomaly that would help in locating the deposit of a particular mineral or to establish the probable sequential extension of a deposit located earlier.[6-8]

Baseline geochemical maps that depict the regional spatial variation of trace elements such as Ag, As, Au, Bi, Ca, Co, Cr, Cu, Ni, Pb, Pd, Pt, Rn, Sb, Si, Te, Zn F, Fe, He, K, Mn, Mo and Na are generated by the multi-elemental study of soils and rocks over the earth surface at spacing intervals on a scale of tens to thousands of kilometre. The purpose of the exercise is to aid in future mineral searches and for decisions on general environmental issues.

Magnetic surveys are a geophysical prospecting method to map or image anomalies in the earth's magnetic field arising from changes in structure or magnetic

FIGURE 1.1 The use of a magnetometer for geological surveying on the surface.

Source: **Courtesy of Wikimedia (2022a)**

susceptibility in certain near-surface rocks. Most magnetic prospecting is now carried out with airborne instruments. In seismic prospecting, vibrations are developed using small explosive charges detonated on the ground or by using some other artificial sources. The ensuing waves of the vibrations that undergo sub-soil travel are precisely measured and from the results, the extent of underlying strata in depth or structural variations is revealed in the image produced. The seismic method can be of reflection or refraction type. Figure 1.1 shows the use of a magnetometer for ground survey.[9]

In an airborne geological survey, an aircraft flying at a low height of 60 m is used to collect geophysical information on the properties of soils, rocks and waters under the ground. At the height of its flight, the aircraft is able to identify geological properties not obtainable from conventional mapping techniques. Airborne survey effectively exposes features like peat and soil cover as well as glacial deposits. It has been found to be the most rapid and cost-effective way to acquire regional-scale geophysical data to support geological mapping, mineral exploration, geothermal energy exploration and radon risk mapping. For a typical airborne survey, the aircraft is equipped with three geophysical instruments which determine the magnetic susceptibility, radioactive level and magnitude of conductivity of the earth below. These measurements are carried out as the aircraft fly at a speed of about 216 km/h, a height of 60 m and at a spacing between pairs of aircraft of 200 m.

1.8   The Earth's Core and Ore Mineral Formation

The earth was presumed formed at very high temperatures and this implies that most metals would have been chemically un-combined at such temperatures notwithstanding the abundance of oxygen. The un-combined heavier metals, such as iron might have then been pulled by gravity inward to the centre resulting in the formation of a metallic magma core. The surrounding envelope of the earth's core consisting of light elements such as oxygen, silicon and aluminium, at lower temperatures might have formed a semi-molten magma consisting of silica and silicates about 3218 km thick. The semi-molten magma subsequently crystallized near the surface to form the solid earth's crust.

The earth's crust is about 37–48.3 km in depth on the continental shelf but thinner under the ocean. The upper layer of the earth's crust consists mainly of granite which is made up of aluminium and calcium silicates and a lower layer of basalt, which comprises mainly magnesium and iron silicates and other mixed silicates such as garnet. As the magma cooled at its surface layer, the most refractory compounds in it solidified first because of their higher solidification point. Heavy compounds, such as those of chromium then sank down into the magma layer too deep to be mined forming original primary deposits. The development of high pressures in the magma and other processes occasionally forced heavy compounds and the lower layers of magma up through crevices to form injection deposits such as Canadian Sudbury copper and nickel-rich ores. The formation of mountains by the folding of the crust caused the enclosing of mineral deposits while the subsequent erosion of these mountains by chemical and mechanical weathering led to the uncovering of layers that were originally deep inside.

During the cooling of the high-temperature region of the earth's crust, most of the elements in dilute solution tend to concentrate on the last traces of the liquid that remained. At about 600°C, the liquid remained solidified to form the pegmatite rocks that host many metallic ores. Escaping gases also forced part of the liquid that remained up into fissures and produced rich vertical veins of ores through the silicate rock. The metallic ores of copper, silver, gold, lead, zinc, cobalt, arsenic, antimony, bismuth and mercury are examples of such ores. When water descending from the surface comes in contact with such deposits mentioned, hot water solutions which later crystallized out in cooler regions were formed to produce hydrothermal deposits such as pitchblende.[8]

1.9   Formation of Accessible Ore Deposits

The earth's crust consists of four main types of igneous rocks. These are rhyolite, andesite, dacite and basalt and they contain twelve oxides. The three main oxides in rhyolite are silica, alumina and potassium oxide at 73.66%, 13.45% and 5.35%, respectively, while andesite has silica, alumina and calcium oxide at 54.20%, 17.17% and 7.92%, respectively. Furthermore, dacite contains silica, alumina and calcium oxide at 63.58%, 16.67% and 5.53%, respectively. Similarly, basalt contains silica, alumina and calcium oxide at 50.83%, 14.07% and 10.42%, respectively. It can be seen that silica is the most dominant oxide in rocks and will thus predominate as a gangue in ore minerals. An ore is a geological anomaly and can be described as an aggregate of minerals from which one or more metals can be profitably extracted. An ore mineral consists of several mineral components mechanically mixed together.[10]

1.10 Mechanisms of Accessible Ore Deposit Formation

Some natural and unusual processes remove minerals, specific elements or compounds from their original primary, in-accessible ordinary rock locations, transport and concentrate them by preference at a spot or zone where the transport stops to form accessible mineral deposits. The unusual process may be:

1. Physical or chemical weathering of mineral hosting original rocks to liberate the mineral values and are thus made ready for transport by rainwater or wind.
2. Sorting by the density of the eroded rock-hosted minerals, the usually heavier mineral values and the lighter gangue minerals.
3. Sorting by the solubility of eroded rocks because some may be soluble in water while others remain insoluble.

1.11 Formation of Ore Deposits

Generally, an ore deposit may be either of igneous, sedimentary or metamorphic origin. Igneous ore deposits are formed when magma originating from the earth's core crystallized into solids in different environments. Apart from ore minerals accumulated and enclosed in rocks, called country or host rocks, mineral deposits may also be found existing on their own in layers and bodies of various shapes which extend over many kilometres in area coverage and of many metres deep. Examples of such independent deposits include deposits of iron oxides and hydrous aluminium oxide in sediments.

For ore deposits in rocks, the first stage of mining involves the removal of the host rock, while the ore minerals are afterwards separated in the second stage of mining. Ore deposits enclosed in host rocks are divided into two groups, syngenetic and epigenetic, on the basis of the time of their formation in relation to their host rocks. Syngenetic deposits are of either igneous or sedimentary origin, and their formation occurred simultaneously with those of the rocks that enclosed them. On the other hand, epigenetic deposits are those deposits that were formed after the host rocks were formed. They are either of igneous, sedimentary or metamorphic origins.

Ore minerals are also classified as hypogene or primary deposits and supergene or secondary deposits such that the secondary deposits are formed when primary deposits are altered through the natural processes of weathering and others related to it. Ore deposits of any of these three broad types of origin and of economic value may owe their origin to a set of simple or complex processes described as follows.

1.12   Igneous Deposits

The main types of igneous deposits are magmatic, pegmatic and hydrothermal deposits.

1.12.1   Magmatic Deposits

Based on the initial composition of molten magma, minerals of economic value are formed when it crystallizes. The value minerals formed may get accumulated at the bottom or margins or other places within the cooling magma due to some factors when the crystallization occurs and may be in such size and nature considered economic for mining. Deposits of magmatic types are known as the main sources of some important metals like gold, uranium and titanium. The accumulation of these economic deposits may occur at an early or later stage during crystallization and is thus known as early or late magmatic crystallization. Magmatic ore deposits may also commonly occur as either segregations, disseminations or injections.

1.12.2   Segregations

In magmatic segregations, some economic minerals like that of chromium get concentrated in the earlier stages of crystallization, when most of the magmatic melt is still in a molten state and such minerals sink to and settle at the bottom due to gravity such that this natural process is called gravitative settling. Similarly, towards the latter stage of magmatic crystallization, when most of the magma has crystallized into solid rock, the remaining melt that contains some economic minerals such as that of titanium get crystallized along the margins of the magma already solidified. This natural process is called residual magmatic segregation.

1.12.3   Disseminations

In this case, economic minerals are dispersed randomly in an igneous rock formed when the molten magma cooled. In some instances, economic minerals are dispersed over a limited space that makes them economically recoverable, while in others the dispersion occurs over a large volume making recovery uneconomical. However, when the latter is the case, but the trend is common over large areas, the whole of the rock mass is considered economic for mining. For example, diamond disseminations are in the famous diamond pipes of South Africa as dispersed crystals in ultrabasic rocks called kimberlites.

1.12.4   Magmatic Injections

Injection deposits are typically thin, slightly tubular or lenticular and may occur at early or late stages of magma crystallization. They are formed when magma erupts with hydrostatic forces that are sufficiently high and get intruded into available fissures as well as fractures in the surrounding rocks where they are eventually cooled.

1.13   Specific Ore Formations

Examples of how some specific ores are formed are given below

1.13.1   Gold Nuggets Formation

Natural occurrence of a gold vein in a pebble or quartz weathered and carried into a river bed. The brittle quartz is subjected to mechanical weathering through abrasion by high-velocity water which chips away the quartz while the malleable gold may be deformed but not reduced. A nugget of gold is eventually formed. Figure 1.2 shows a gold nugget.[11]

1.13.2   Formation of Spinel Group of Minerals

In comparison with other minerals, the spinel group of minerals exhibit high density and high melting temperatures. The typical chemical formula for spinel is $MgAl_2O_4$ where

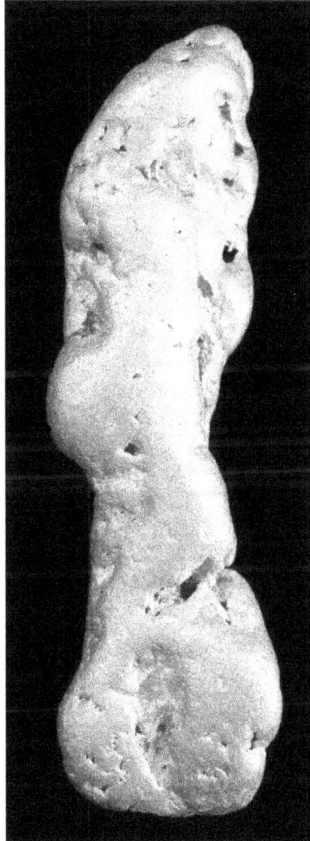

FIGURE 1.2   Gold nugget.

Source: Courtesy of Wikimedia (2022b)

FIGURE 1.3   Spinel mineral sitting on calcite.

Source: Courtesy of **Wikimedia (2022c)**

Mg^{2+} can be substituted with Fe^{2+}, Zn^{2+}, Mn^{2+}
Al_3^{+} can be substituted with Fe^{3+}, Mn^{3+}, Cr^{3+}
Ti^{4+} and V^{3+} can also get substituted into the structure

Figure 1.3 shows a spinel mineral sitting on calcite.[12]

Basaltic magmas in the magma chamber typically contain about 10% Fe, a few per cent Ti and trace amounts of Cr, Mn and V and have a density of 2500 kg/m^3. When the right conditions are obtained, the elements they contain become concentrated as spinel minerals with a high density of 4800 kg/m^3. The high-density spinels will sink into the bottom of the chamber forming layers of spinel groups of minerals of different specific gravities.

1.13.3   Magmatic Mineral Deposit by Crystal Settling

When we have molten magma, crystals of minerals such as olivine, chromite and plagioclase can crystallize in it. The grains of these minerals will have different

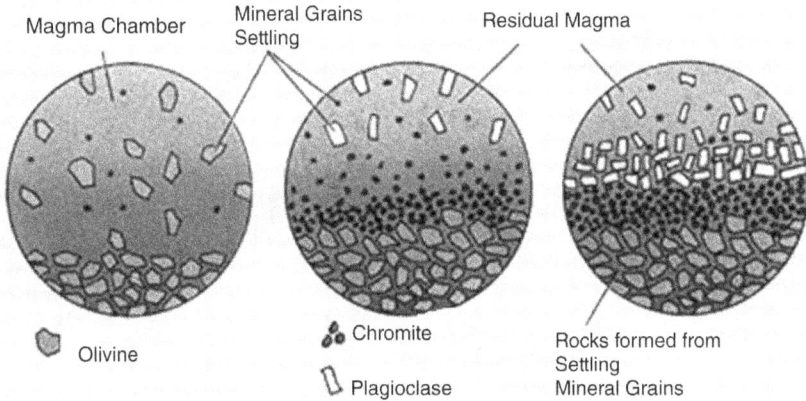

FIGURE 1.4   Grains of olivine, chromite and plagioclase with different settling rates.

Source: Skinner and Stephen (1995) (courtesy of John Wiley & Sons, Inc., adapted)

settling rates and will form a rock with layers of olivine, chromite and plagioclase as shown in Figure 1.4.[10]

1.13.4   Ore Formation from Magmas by Liquid Immiscibility

When the high-temperature magma cools, it gets transformed into two liquids having different volumes, densities and compositions at a lower temperature. The higher volume liquid is high in silica content while the second liquid is richly populated with grains of metal sulphides, oxides or carbonates which settle and get concentrated in it.

The melt rich in sulphides, oxide or carbonates has the following features:

a. The oxide melts may contain a high quantity of Fe as Fe_2O_3, hematite and $Ti(FeTiO)_3$, ilmenite.
b. The sulphide melts may be richly populated with Cu, Ni, the platinum group metals (PGMs) and alloys of iron and sulphur, that is, FeS and pyrrhotite.
c. The carbonate melts may be richly endowed with metals such as niobium, tantalum, rare earth, copper, thorium and phosphorous.

1.13.5   Pegmatite Deposits

Pegmatite rocks are sources of mica, quartz and many strategic minerals such as beryl, lithium minerals, rare earths and numerous gems. They are magmatic deposits formed towards the end of the crystallization process and as a result, occur close to magmatic masses roofs. The pegmatites may be either of simple type or of complex type. Pegmatites may occur in the forms of dikes, veins, lenses and nests of variable sizes and dimensions. Some pegmatites are characterized by exceptionally large-sized crystals. For instance, the crystals of the pegmatic beryl mineral can be as large as 6 m in length and 1.3 m in diameter and can weigh up to 18 tons as found in Maine, in Albany, United States. Pegmatites are often associated with granites and are called granite-pegmatite.

1.13.6   Hydrothermal Deposits

These economic minerals are formed when a solution containing some metal values exists in the form of a superheated steam or liquid gas coming from magmas towards the end of crystallizations enters and gets cooled in cavities, fissures or pores of rocks. The superheated steam can contain metallic elements like gold, copper, tungsten, molybdenum and to some extent silver, lead and zinc in dissolved form. The deposits may form due to changes in temperature or pressure or both or as a result of chemical reactions between the steam's various components or because of chemical reactions between the steam components and the country rock with which they interact. These dissolved elements get crystallized out as the carrier-enriched solutions move upwards and experience cooling and loss of pressure. They may occur as veins and cavity fillings. Veins are narrow, elongated or tabular bodies of economic minerals occurring within a host rock of entirely different composition and origin.

Metasomatic (replacement) deposits are formed from pure hydrothermal (magmatic) deposits due to the process of replacement of some original components by a new component in a solid state change. The replacement deposits are only commonly formed in rocks of suitable composition which belong to the carbonate group, that is, limestones and dolomites. Igneous rocks have also been found altered by the replacement (metasomatic) process. They contain metals like beryllium, caesium, lithium, niobium, rubidium, tantalum, thorium, tin, tungsten and uranium; and non-metals like quartz and fluorite; as well as gemstones like amethyst, aquamarine, garnet and topaz. The replacement deposits may be found as veins, lodes or zones. A typical example is the porphyry copper-molybdenum deposit.

The large magma volume that formed the porphyritic copper bodies has its origin several kilometres below the location of the deposit. The deposit derives its name from the vertical dikes of porphyritic intrusive rocks with which the magma fluids are associated.

1.13.7   Marine Hydrothermal Deposits

They are commonly found as mid-ocean ridges. The mechanism of the formation is as follows:

a. Oxides and sulphides in hot basaltic crust rock at temperatures above 300°C are selectively dissolved by water that percolates through it.
b. The water-bearing dissolved minerals are discharged into the ocean and the dissolved minerals precipitate as soon as the hot water cools.
c. This process can cause the concentration of metals such as copper, lead, zinc and silver as volcanogenic sulphide deposits of massive size.
d. Since the different minerals precipitate at different temperatures, their precipitation will occur at different places on the ocean floor.

1.13.8   Sedimentary Deposits

Some ore deposits of iron, copper, gold, phosphates and coal are of sedimentary origin. The accumulation of sedimentary deposits occurs according to the following processes.

1.13.8.1   Weathering

Gases from the atmosphere and water vapour are constantly reacting with the surface rocks and rocks of suitable composition that get decomposed, disintegrated and become altered. The disintegrated and altered rock product is called a mantle of waste and constitutes the source material from which ore deposits of economic value are formed. The gangue minerals in the weathered rocks are leached by rain and subsurface water leaving residual material of economic value. Ores of iron such as goethite and limonite as well as nickel silicate deposits are residual deposits. The soluble values in the disintegrated rocks are dissolved by water and get re-deposited at lower altitudes when the carrier solution infiltrates the lower rocks and are called infiltration-type deposits. Ores of uranium, some bauxite types and silicate-carbonate iron ore groups are examples of infiltration-type deposits. Figure 1.5 shows a bauxite ore lump formed by weathering.[13]

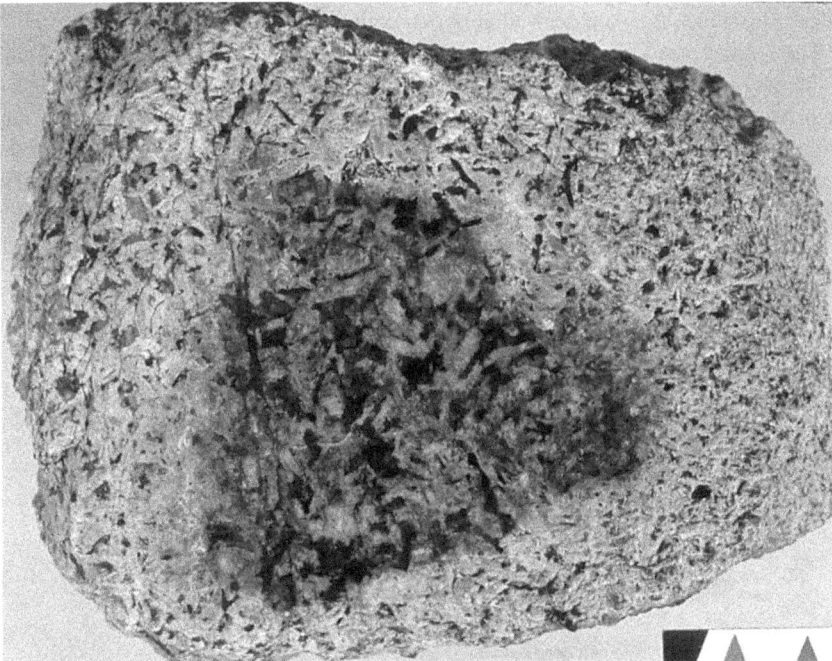

FIGURE 1.5   Bauxite with a core of unweathered rock.

Source: Courtesy of Wikimedia (2022d)

1.13.8.2   Sedimentary Genesis

These deposits are formed when rocks disintegrate physically and chemically by weathering and the mantle of waste is washed by erosion into sediments. The sediments are transported by water, wind and ice to environments suitable for deposition. The sediment deposition is then followed by diagenesis of the sediments to sedimentary formations. The deposits of pure sedimentary origin occur very commonly in the form of layered formations of extensive areal extent and considerable depth. In economic value, they range from non-metallic deposits like sandstones and limestone, clays, salts or sodium, potassium and magnesium to those of important metals like iron, manganese, aluminium, copper, uranium and vanadium. Coals also form a distinct group of sedimentary deposits in which biochemical changes have also taken place during the process of deposition and diagenesis. Figure 1.6 shows reddish-brown a bauxite of sedimentary origin.[14]

1.13.8.3   Placer Deposits

Placer deposits are a special group of deposits of weathering origin that have been released from their original rock bodies by weathering. They were transported by natural actions of wind, water, ice and gravity and got accumulated in varying locations because of their unique properties such as high density, chemical stability, hardness and resistance to abrasion. The placer deposits are called deluvial, alluvial, aeolian and beach placers if the causal agent for the transport and deposition is gravity, running water, wind and waves, respectively. Placer deposits found accumulated on the rocks from which they are derived are often called elluvial placers.

The formation of placer deposits at their locations depends mainly on the ability of the carrier medium to sort the materials and the inherent physical properties of the ore mineral concerned. For example, high density, hard and stable minerals have better chances of getting accumulated and being preserved as deposits compared to lower density and chemically unstable minerals that are subjected to further disintegration and dispersion. The placer deposit mechanism leads to the formation of ores of gold, diamond, platinum, tin, tungsten, zircon and minerals like magnetite, monazite and garnet.

FIGURE 1.6   Reddish brown sedimentary type bauxite.

Source: Courtesy of Wikimedia (2022e)

Heavy mineral particles found on the stream beds are placer ore deposits. They are formed when mechanical and physical weathering removes mineral particles from country rock and streams at high velocity carry the mineral particles removed by weathering. At points where the kinetic energy of the stream reduces due to the stream's encounter with rock bars, meander loops, rock holes, waterfalls or downstream from a tributary, high-density particles will stop while low-density gangue particles of quartz will continue moving. For this to happen, the density contrast of the heavy mineral with quartz must be high as in the case of gold and quartz with densities of 19 and 2.65 g/cm^3, respectively. Panning uses the principles of placer deposit formation to separate lighter particles from a sample to retain the heavy gold particles. Figure 1.7 shows the mechanisms of placer deposit formation.[10]

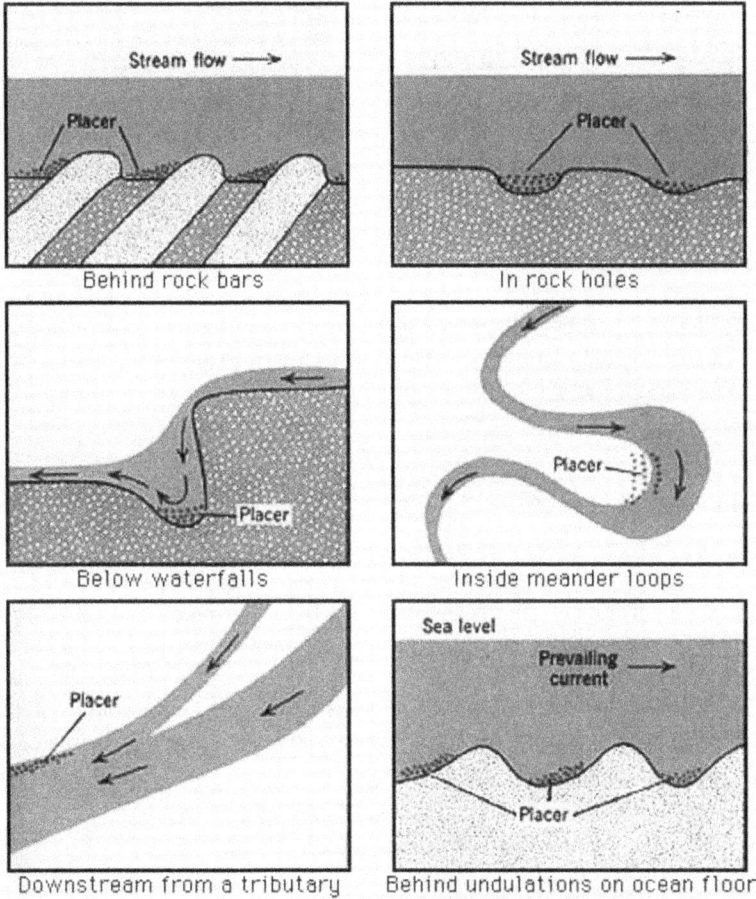

FIGURE 1.7   Formation of placer deposits.

Source: **Skinner and Stephen (1995) (courtesy of John Wiley & Sons, Inc., adapted)**

Placer deposits that are of economic importance include nuggets of gold, silver, platinum, diamonds (carbon), zircon (zirconium silicate), uraninite (uranium oxide) and rutile (titanium oxide).

1.13.9 Metamorphic Deposits

Metamorphic deposits are formed from pre-existing sedimentary and igneous rocks that have experienced changed conditions of temperature and pressure and are in contact with chemically active fluids resulting in the formation of new minerals of economic value. For instance, limestone is metamorphosed to marble and slate deposits are formed from shales by metamorphic transformation. Generally, metamorphism affects non-metallic minerals producing new non-metallic minerals of higher economic value.[15]

1.14 Ore Formation Mechanisms that Involve Oxidation State in Water

Dissolved mineral materials containing metals with variable oxidation states can be dissolved in groundwater and carried along in its flow. The dissolved metals can get precipitated out of the groundwater solution if the water oxidizing conditions increase or decrease as the water flows. For example, uranium exists in two stable oxidation states, that is, the soluble U^{6+} and the insoluble U^{4+} which is produced when igneous rocks are weathered.

If uranium in the insoluble state U^{4+} gets in contact with oxidizing water, it becomes oxidized to the soluble U^{6+} and gets transported by groundwater. If the water-bearing soluble uranium encounters a reducing condition such as contact with buried wood, the soluble uranium gets reduced to U^{4+} to form the yellowish uranium oxide.

1.15 Spread of Solid Minerals in Nigeria

The occurrence of minerals is related to certain geological anomalies. A geological anomaly is a geologic abnormality exhibiting features that depart markedly from its surrounding environment. In areas with geophysical anomalies, geophysical properties such as radiometry, magnetism, electromagnetism and response to gravity differ from surrounding areas. Geochemical anomaly refers to the concentration or accumulation of one or more elements in a rock, soil, sediment, vegetation or water that is markedly higher or lower than the general background material. The mineral deposits of Nigeria are found as follows.

1.15.1 The Benue Trough of Nigeria

The Benue Trough is an elongated rifted sedimentary basin that stretches from the Gulf of Guinea to Lake Chad over a distance of about 1000 km.

It is characterized by sediments several metres thick and cretaceous rocks about 480,000 km^2 in area. It is reported to contain about 25 million metric tons of lead,

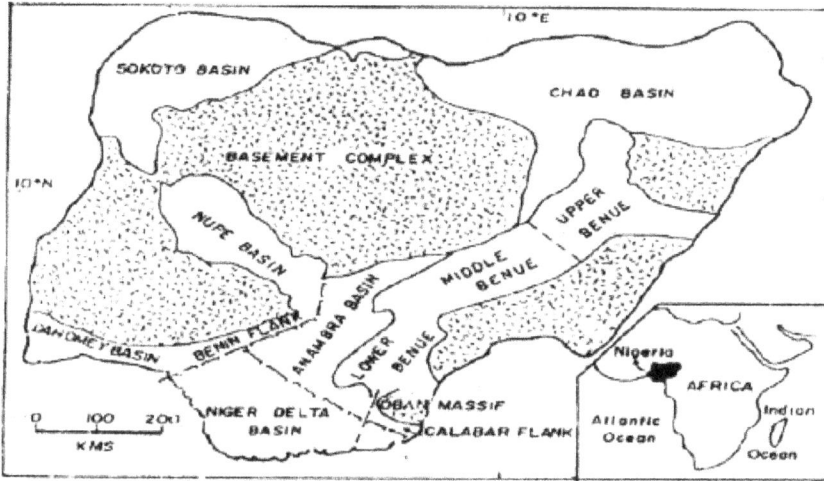

FIGURE 1.8   Map of Nigeria showing the location of the Benue Trough (Basin) and South—Eastern Sedimentary basins (Akana and Didei, 2017).

zinc and copper. It is subdivided into the lower, middle and upper Benue Troughs. The lead-zinc deposits of Zurak in Plateau state and the Akiri copper deposits of Nassarawa state are found in the middle Benue trough, while the lead-zinc deposits of Ishiagu in Ebonyi state are found in the Lower Benue trough. Figure 1.8 is a map of Nigeria showing the location of the Benue Trough (Basin) and South—Eastern Sedimentary basins.[16]

1.15.2   Sedimentary Rock Iron Deposits

They are associated with sedimentary rock and are formed either by physical or chemical sedimentation. The deposits due to physical sedimentation are called banded iron formations while the ones due to chemical sedimentation are called oolithic iron deposits. Oolite is a sedimentary rock formed from ooids which are spherical grains composed of concentric layers. In physical sedimentation, solid particles entrained in the turbulent flow of water in motion may be deposited naturally by sedimentation mechanism in the stationary waters of lakes and oceans. Chemical sedimentary rocks form when dissolved materials precipitate from a solution. Examples of mineral deposits found as sedimentary rocks include iron ore, dolomite, rock salt, flints and limestones.

Itakpe, Chokochoko and Ajabanoko iron ore deposits are examples of banded iron formations, while oolithic iron ores at Agbaja and Kotonkarfe deposits in Kogi state are typical examples of sedimentary iron ore deposits.

1.15.3   The Minor Sedimentary Basin

This basin comprises the Benue, Kogi and Enugu districts spanning 1.5 million hectares and hosting about 396 million tons of coal.

1.15.4   Magmatic Ore Deposits

A magmatic ore deposit is an accumulation of magmatic minerals accumulated in large amounts and at un-usually high metal concentrations. The younger granite of Plateau state contains cassiterite ore in association with other minerals such as rutile, magnetite, columbite, tantalite, zircon, monazite and wolframite. However, some magmatic deposits contain extremely rare minerals such as PGMs that are almost never encountered in common rocks.

1.15.5   The Metamorphic Rock Deposits

Common metamorphic rocks include schist, gneiss, quartzite, phyllite, quartzite and marble. The schist is a metamorphic rock of the medium-grade category. It consists of medium to large, flat, sheet-like grains oriented in a preferred direction. The supra-crustal schist belt of southwest and northwest of Nigeria contains gold found in places like Iperindo, Itagunmodi, Maru, Malele and Birnin Gwari.

1.15.6   The Carbonate Province

The carbonate provinces of Nigeria include Ewekoro, Nkalagu, Sokoto, Yandev and Ashaka. The Mfamosing deposit in Cross River state contains about 97% $CaCO_3$.

1.15.7   Coastal Deposits

The coastal deposits are found at the ocean beaches and include quartz sand such as the one found in Trans Amadi, Port Harcourt with about 97% silica. Rutile is also found in some coastal beach sands.

1.15.8   Erosion Deposits

Chemical and mechanical weathering of rocks occurs. When such rocks host heavy minerals like cassiterite or heavy metals like gold, they are carried along river beds and deposited. Such deposits are called placer deposits. Erosion deposits of cassiterite and gold are found in Nigeria.

In fulfilling its mandate, the Nigeria Federal Ministry of Mines and Steel Development designated seven notable minerals, barite, gold, bitumen, iron ore, lead/zinc, coal and limestone as strategic minerals because they are believed to have the potential to make significant contributions to national economic development.

1.16   Notable Mineral Resources of Africa

Africa is richly endowed with various mineral deposits and it has been reported to hold about one-third of the world's mineral deposits. The United States Geological Survey (USGS) has reported that Africa has large reserves of ores of platinum, diamonds, bauxite, gold, manganese, cobalt and others. It has been noted that South Africa has the world's largest reserves of platinum group metals (PGMs) and chromite ores in its Bushveld Igneous Complex deposits. The Great Dyke of Zimbabwe also holds deposits of PGMs. Diamonds are found all over the continent and copper occurs in large deposits in the Democratic Republic of Congo (DRC), Zambia, Tanzania, South Africa and Namibia.

The Bushveld Igneous Complex is a layered complex of igneous rocks, about 9 km thick and covers an area of about 66,000 km^2. It is the world's largest reserve of magmatic ore deposits such as chromites. It also hosts the world's largest deposits of PGM ores. The Great Dyke of Zimbabwe, the Sudbury deposit of Ontario, the Norilsk Talnakh of Siberia in Russia and the Stillwater Complex in the United States also host PGM ores in lesser reserves. The section refers to black chromite and grey anorthosite layered igneous rocks in the Critical Zone UG1 of the Bushveld Igneous Complex at the Monanono River outcrop, near Steelpoort, Limpopo province, South Africa, which is. part of the Winterveld Norite-Anorthosite unit of the Dwars River Subsuite of the Rustenburg Layered Suite.[17] Figure 1.9 shows the chromite layers of the Bushveld Igneous Complex.[18]

Niger is endowed with two notable high-grade uranium deposits at Azelok and in sandstones such as those of Abokurum, Arlette, Imouraren and Tassa. In 2019, Niger produced 2,983 tons of uranium. The Imouraren deposit estimated to contain 179,000 tons of uranium is the largest uranium deposit in Africa and the second largest in the world.

FIGURE 1.9   Chromite layers of the Bushveld Igneous Complex.

Source: Courtesy of Wikimedia (2022f)

Guinea Conakry has a large deposit of diamonds at Banankoro in the Eastern part of the country estimated to be 30–40 million carats of proven reserves. Ghana has large reserves of alluvial gold deposits found in small streams and rivers flowing over primary rock deposits, particularly around the Atewe Range of the country. It was reported that over 267 kg of gold was produced from about 298,000 cm^3 of gravel whose concentrates assay 0.8–0.9 g/cm^3.

Zimbabwe has a large and diverse solid mineral resource found in its two main anomalous geological features, that is, the great Dyke and the Greenstone or Gold belts. The great Dyke is a complex layer of igneous rocks of about 550 km in length that extends from the North to the South of the country. It holds the world's largest reserve of chromite estimated at 10 billion tons which constitutes about 80% of the world's reserves and the second largest reserve of platinum group metals ores estimated as 2.8 billion tons at an average grade of 4 g/t as well as significant quantities of copper and nickel. The Greenstone belts also hold more than 90% of the gold deposits of Zimbabwe. The country is also endowed with ancient basins called cratons that hold diamond deposits. Zimbabwe also has over 12 billion tons of high-grade coals found in 29 deposits suitable for thermal and metallurgical applications. Furthermore, there is a large deposit of iron ore from the banded ironstone formations estimated at about 30 billion tons in the Greenstone belt.

Zambia is ranked as the seventh largest copper producer in the world and its copper deposits belong to the category with the world's highest grade. In addition, Zambia produces about 20 per cent of the emeralds' outputs in the world. Botswana has the world's richest diamond deposit in its Jwaneng open pit mines and 70–80% of its foreign exchange income comes from diamond. The country also has untapped mineral resources, such as gypsum, iron ore, asbestos, feldspar, chromite, graphite, and manganese ores found in remote areas of the Kalahari deserts. The country's output of gold was 1,800 kg in 2010.

The Democratic Republic of Congo (DRC) is regarded as the wealthiest nation in the world as it has numerous solid mineral resources including unexploited resources estimated at about $24 trillion which equals the combined Gross Domestic Product of Europe and the United States. It holds the world's largest deposits of coltan, the ore mineral of niobium and tantalum and it accounts for half of the world's resource of cobalt. Furthermore, it has deposits of lithium, tin, tungsten, gold and diamond. Rwanda, another African country is endowed with several solid minerals such as those bearing rare earth elements, gemstones, cobalt, iron and lithium.

Rwanda is also endowed with a pegmatite hosting Li–Cs–Ta (LCT) in the Ruhanga area of the country. As of 2018, a pegmatite was being explored for columbite-tantalite mineralization. The pegmatite body's width was determined to be between 15 and 20 m and it has a lateral extension of at least 2.5 km. The feldspars were completely kaolinized leading to an un-zoned appearance which is characteristic of most of the pegmatites in Rwanda. The pegmatites are also characterized by rafts of biotite schist. Table 1.1 presents some notable ore deposits in Africa.[19–24]

FIGURE 1.10 The earth's crust structures exposed in a Rwandan mine.[25]

TABLE 1.1
Selected Ore Deposits of Africa in Summary

S/N	Mineral	Percentage of World's Output	Countries' Contribution
1	Diamond	73	Botswana (35%), Congo (Kinshasa) (34%), South Africa (17%), Angola (8%)
2	Gold	89	South Africa (56%), Ghana (13%), Tanzania (10%), Mali (8%)
3	Uranium	16	Namibia (46), Niger (44%), South Africa (<10%)
4	Bauxite	9	Guinea (95%), Ghana (5%)
5	Platinum	92	South Africa (96/97%)

1.17 Notable Mineral Resources of the United States

The United States hosts large reserves of non-metallic ore deposits such as limestone and coal as well as ores of iron, copper and many other metals. In 2019, the US outputs of top mineral-based commodities are shown in Table 1.2.[26]

TABLE 1.2
Top Mineral-Based Commodities Outputs in the United States in 2019

S/N	Mineral Based Commodity	Output Value ($US billions)
1	Coal	25.1
2	Crushed rock	18.7
3	Gold	9.0
4	Copper	7.9
5	Iron ore	5.4
6	Industrial sand and gravel	5.7
7	Construction sand and gravel	9.0
8	Cement	12.5

1.18   Notable Ore Deposits of India

India has been reported to host seventeen major minerals. These are bauxite with probable reserves of 90 million tons and having the main reserves on the East coast. The country is also richly endowed with chromite with a geological reserve estimated at 187 million tons. In-situ reserves of iron ore have been put at 1.23 billion tons, while copper and gold have in-situ reserves of 712.5 and 22.4 million tons, respectively. Other solid minerals with major occurrences are those of manganese, lead-zinc, tungsten and industrial non-metallic minerals such as diamond, dolomite, gypsum, fluorspar and graphite estimated at 26.4 million carats, 7.3 billion tons, 383 million tons, 14.2 million tons and 16 million tons, respectively. The rock types of India and the ores associated with them are tabulated in Table 1.3.[27, 28]

1.19   Notable Ore Deposits in Canada

Canada is also richly endowed with solid mineral resources. For thirteen major minerals output, Canada has been placed in the fifth position of lead-producing nations. It takes the lead position in the mining of potash, second position in the production of uranium and niobium, third position in the output of nickel, cobalt, aluminium and PGMs as well as the fifth position in the mining of gold and diamond. In 2016, the minerals and metals industry contributed $87 billion to the national Gross Domestic Product (GDP).[29]

1.20   Introduction to Ore Processing

The mineral processing technologies generally involve drilling, comminution, screening, slurry processing as well as thickening and clarification. Drilling and blasting are used to fragment in situ minerals from their beds to produce lumps usually as large as 1.5 m in size while crushing is carried out dry and grinding wet for ores in the range of about 5 mm to over 1 m. The processing of slurries involves the

TABLE 1.3
Selected Rock Types of India and Associated Mineral Deposits

S/N	Rock type	Associated Ores
1	Kimberlite and Lamproite	Diamond
2	Dunite-Peridotite, pyroxenite	Nickel, chromite, PGMs
3	Norite-Gabbro, Anorthosite	PGMs, magnetite bearing Ti and V, native copper, silver, cobalt, nickel
4	Dolerite, diorite, monzonite	Magnetite, copper, gold
5	Granodiorite, quartz monzonite	Porphyry copper, Au-
6	Syenite	Magnetite, gold
7	Nephene syenite	Corundum
8	Granite and granite pegmatite	Sn-W, uranium, radium, beryl, tourmaline

use of pumps, agitators, classifiers and separators. The slurry processing is followed by product handling which involves the use of thickeners and clarifiers to produce products for subsequent drying, calcining and sintering. The mineral processing operation is also categorized to involve front-end operations, size reductions, enrichment, upgrading, material handling and protection.

Enrichment may entail washing the ore and/or subjecting it to a separation process, while upgrading also refers to processing the ore by separation to obtain concentrates of the desired grade. An ore is washed to remove surface impurities such as dirts, clays, salts or organic materials on the ore using wet screens, scrubbers, attrition cells and gravity beds. The separation takes place by gravity, magnetic, flotation and leaching. The concentrate stream is upgraded by thickening and pyrometallurgy methods of drying, calcining, roasting and sintering. Material handling refers to moving the material through the processing with it being minimally disturbed. Protection refers to measures put in place to protect the processing environment from wear and dust.[2]

References

1. Metso (2022): *Basics in Mineral Processing* (https://vdocument.in/basics-in-minerals-processingmetso.html, Accessed 23rd June, 2022).
2. Frye, K. (1988): Mineraloids. In: *General Geology: Encyclopedia of Earth Science*. Boston, MA: Springer.
3. Wills, B.A., and Napier-Munn, T.J. (2006): *Mineral Processing Technology*, 7th Edition. Amsterdam: Elsevier Science and Technology Books.
4. Hale, M. (1998): Ore Deposits. In: *Geochemistry: Encyclopedia of Earth Science*. Dordrecht: Springer (Accessed 31st May, 2020).
5. Australian Industry and Skills Committee (2022): *Metalliferous Mining (AISC)* (https://nationalindustryinsights.aisc.net.au/industries/mining-drilling-and-civil-infrastructure/metalliferous-mining, Accessed 31st May, 2020).
6. Gandhi, S.M., and Sarkar, B.C. (2016): *Essentials of Mineral Exploration and Evaluation Geological Survey Ireland (GSI)* (www.gsi.ie, Accessed 23rd May, 2020).
7. Swapan, K.H. (2018): *Mineral Exploration-Principles and Applications*, 2nd Edition. Amsterdam: Elsevier.

8. Cottrel, A. (1980): *Introduction to Metallurgy*, ELBS Edition. London: Edward Arnold.
9. Wikimedia (2022a): *Magnetometer* (https://upload.wikimedia.org/wikipedia/commons/thumb/8/87/Pomiary_magnetometrem_transduktorowym.jpg/800px-Pomiary_magnetometrem_transduktorowym.jpg, 110522) By Glab310 — Own Work, CC BY 4.0 (https://commons.wikimedia.org/w/index.php?curid=113524155).
10. Skinner, B.J., and Stephen, C.P. (1995): *The Dynamic Earth*, 3rd Edition. New York: John Wiley & Sons, Inc., p. 109.
11. Wikimedia (2022b): *Gold Nugget* (https://upload.wikimedia.org/wikipedia/commons/9/96/Gold-270415.jpg, Accessed 22nd April, 2022).
12. Wikimedia (2022c): *Calcite-Spinel* (https://upload.wikimedia.org/wikipedia/commons/b/b9/Calcite-Spinel-dtn37a.jpg, Accessed 25th April, 2022); By Rob Lavinsky, iRocks.com—CC-BY-SA-3.0, CC BY-SA 3.0 (https://commons.wikimedia.org/w/index.php?curid=10447446).
13. Wikimedia (2022d): *Bauxite with Core of Un-Weathered Rock* (https://upload.wikimedia.org/wikipedia/commons/thumb/f/f5/Bauxite_with_unweathered_rock_core._C_021.jpg/1024px-Bauxite_with_unweathered_rock_core._C_021.jpg, Accessed 22nd April, 2022) (By Werner Schellmann—Own work, CC BY-SA 2.5 (https://commons.wikimedia.org/w/index.php?curid=1004613).
14. Wikimedia (2022e): *Reddish Brown Bauxite* (https://commons.wikimedia.org/w/index.php?curid=2099323, Accessed 22nd April, 2022) Author from Wiki: By saphon—Own work, CC BY-SA 1.0 (https://commons.wikimedia.org/w/index.php?curid=2099323, 220422).
15. Malik, R. (2021): *Mineral Ore Deposits: Meaning, Origin and Types, Geology* (www.geographynotes.com, Accessed 5th November, 2021).
16. Akana, T.S., and Didei, I.S. (2017): Paleocurrent Analysis of the Sandstone Body in Akpoha and Its Enviorns (Lower Benue Trough) South-East Nigeria. *Journal of Geology & Geophysics*, 6, 309 (https://doi.org/10.4172/2381-8719.1000309 Copyright; file:///C:/user/Downloads/Paleocurrent_Analysis_of_the_Sandstone_Body_in_Akp.pdf, Accessed 16th May, 2022).
17. Britannica (2020): *Bushveld Complex* (www.britannica.com/place/Bushveld-Complex, Accessed 30th August, 2020).
18. Wikimedia (2022f): *Bushveld Igneous Complex* (hhttps://upload.wikimedia.org/wikipedia/commons/thumb/8/81/Chromitite_Bushveld_South_Africa.jpg/450px-Chromitite_Bushveld_South_Africa.jpg, Accessed 20th April, 2022).
19. AZO Mining (2022): *Botswana: Mining, Minerals and Fuel Resources* (htttps://www.azomining.com/Article.aspx?ArticleID=83, Accessed 22nd March, 2022).
20. Imouraren Uranium Mine—Mining Technology, Mining News (https://www.mining-technology.com/projects/imouraren-uranium-mine-niger, Accessed 29th December, 2021).
21. Ministry of Mines and Mining Development, Zimbabwe (https://miningzimbabwe.com/ministry-of-mines-and-mining-development-mmmd, Accessed 22nd March, 2022).
22. Republic of Guinea-Mining and Minerals-International Trade (https://www.trade.gov?guinea-ministry, Accessed 22nd March, 2022).
23. Sande, K. (2021): *An Estimated $24 Trillion in Untapped Deposits of Raw Mineral Ores Lie Beneath the Congo's Incredibly Vast Lands* (https://www.sandekennedy.com/an-estimated-24-trillion-in-untapped-deposits-of-raw-mineral-ores-lie-beneath-the-congos-incredibly-vast-lands/, Accessed 22nd March, 2022).
24. World Nuclear Association (WNA) (2021): *Uranium in Niger Uranium in Niger, World Nuclear Association* (https://www.world-nuclear.org, Accessed 29th December, 2021).

25. Steiner, B.M. (2019): Tools and Workflows for Grassroots Li–Cs–Ta (LCT) Pegmatite Exploration. *Minerals*, 9, 499 (https://doi.org/10.3390/min9080499; www.mdpi.com/minerals/minerals-09-00499/article_deploy/html/images/minerals-09-00499-g001.png, Accessed 11th May, 2022).
26. USGS (2020): *Mineral Commodities Summaries* (https://www.usgs.gov/media/images/mineral-commodities-february-2020, Accessed 30th August, 2020).
27. Geological Survey of India (2022): *Mineral Resources of India (Geology and Mineral Resources . . .* (https://www.eptrienvis.nic.in, Accessed 17th March, 2022).
28. Library.com. *17 Major Mineral Resources Found in India* (https://www.yourarticlelibrary.com, Accessed 21st March, 2022).
29. The Canadian Minerals and Metals Plan (2022): *Mining in Canada (Mining in Canada | Mines Canada)* (https://www.minescanada.ca/en, Accessed 22nd March, 2022).

2 Ore Handling

2.1 Introduction

Ore handling refers to the processes of transporting, storing, feeding and washing the run-of-mine ore on its movement to the mill or within the mill in the course of the various stages of its treatment. For run-of-mine ore (ROM) handling, the cost may translate to between 30% and 60% of the total cost of delivery to the buyer. The physical condition of the ore as found in its deposit, that is, whether it is sandy, friable or monolithic, which means massive, with a hardness similar to that of granite having Mohs hardness of 6–7 will determine the mining methods and the way to handle the freshly mined materials. Generally, the following should be noted about ore handling:

1. Ores that are subjected to proper breaking can be transported on conveyor belts, by trucks or possibly using a sluice box.
2. Individual blasting may be carried out for large lumps of hard ore using delay fuses, plastic explosives such as Semtex and C-4, low-density explosives and chemical explosives such as ANFO, that is Ammonium Nitrate/Fuel Oil (AN) which constitutes 94% and serves as the oxidizing agent plus 6% number 2 fuel oil ((FO). FO is a distillate home heating oil. Plastic explosives are most suited for the explosive demolition of very large lumps.
3. Large primary crushers that can accept lumps that are as large as 2 m in size can be used. There are solidly built modern primary gyratory crushers that are designed for large tonnage outputs and can accept large lumps of run-of-mine rocks directly. The allowable feed size can be defined as 0.9 × gape with a size feed of up to 2 m and reduction ratios ranging from 3:1 to 10:1.
4. Open pit mining is used when the deposits of useful ore are shallow or found near the surface and the equipment required include large excavators, large wheel loaders, large bulldozers, hydraulic mining shovels, electric rope shovels, drills, draglines and trucks that can carry up to 200 tons of ore. Other mining methods are placer, in situ and underground mining.
5. Underground mining is used for ores that are deep in the earth and overlaid by rocks. It is too expensive to remove all the overlain rocks. Tunnels have to be made in the rocks so that miners and equipment can gain access to the ore. A mine shaft is required too. It is a vertical or near-vertical access hole that is several metres in diameter and extends down to the ore location. Ore is hoisted up through it using mine skips that can carry up to 30 tons of ore. Primary breakers that can break rocks to about 18 cm in size after blasting are used to crush large rocks underground.[1]

DOI: 10.1201/9781003323433-2

2.2   Harmful Materials Separation from Ore

The ROM, that is, the ore being delivered to the milling plant from the mine usually contains some materials that can damage mill equipment or adversely affect the
process flow. These harmful materials include the following.

2.2.1   Steel Scrap

Steel scrap broken off mining machineries will get jammed in the crusher if allowed
to enter it. They must therefore be removed using powerful electromagnets suspended
over the conveyor belt transporting the ore.

2.2.2   Wood

Wood in the ore is ground to pulp that can block screens and choke flotation cell ports.
Wood in the slurries also negatively affects the froth flotation process by increasing
flotation reagents consumption through absorption. They are also reported to decom-
pose forming depressants and thus render otherwise floatable valuable minerals un-
floatable. Flattened woods exiting the primary crushers are removed using vibrating
scalping screens with apertures much larger than the crusher product diameter. In
grinding, the ground product discharge can be passed over fine-sized screens to
remove wood pulps.

2.2.3   Clays and Slimes

Clays, dirts and slimes coating the ore particles negatively affect the performance
of screens, filters and thickeners and can increase the consumption of froth flotation
reagents. The ROM is subjected to washing on washing screens after primary crushing to remove adhering dirts and slimes.

2.3   Transportation of Ore

2.3.1   The Challenge of Ore Transport in the Plant

A typical mineral processing plant that treats 14 tons of ore per minute will
require about 37.5 m³ of water. In a day's operation, such a plant will need to
transport over 20, 000 tons of ore material during treatment in the plant and use
54,000 m³ or 54 million litres or 14,210,526 gallons of water per day. This quantity of water is expected to be used by about 375,000 people per day at a rate of
144 litres per person. The basic philosophy of ore transport in a plant, therefore,
needs to:

  a. Reduce energy usage by minimizing horizontal and vertical movement in
     the plant
  b. Maximize sloping movement under gravity fall

c. Reduce water consumption by using slurries with as high pulp density as much as possible

d. Design distances between unit operations to be the shortest possible length

The ore material to be transported may be dry solids, loose bulk material, wet sand, sticky materials and wet slurries. Dry ores are transported on steel-faced feed chutes with 25–45° slopes and 3–4 ft long, while loose bulk ores are moved over rubber conveyor belts with a speed that can be as high as 10 m/s. Wet sand and sticky materials are also moved with chutes. Wet slurries are transported in pipelines under a centrifugal pump or displacement pressure. From the grinding stage onward, dry transport of ore is replaced by hydraulic transport in pipelines and launders. Figures 2.1, 2.2 and 2.3 show a chute, Eriez magnetic belt conveyor and Eriez industrial ore conveying system, respectively.[2–4]

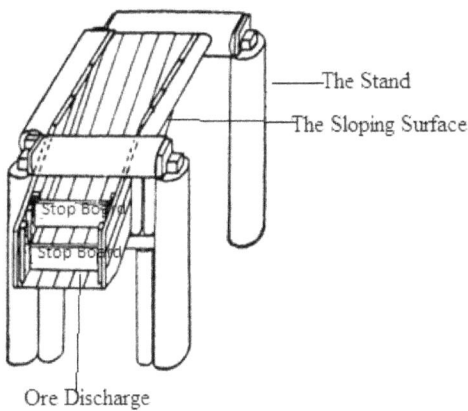

FIGURE 2.1   A chute.

Source: Courtesy of 911 Metallurgist (2022a)

FIGURE 2.2   Eriez magnetic belt conveyor.

Source: Courtesy of Eriez (2022a)

FIGURE 2.3   Eriez industrial ore conveying system.

Source: **Courtesy of Eriez (2022b)**

2.4   Ore Storage

Storage can be done in bins, stockpiles or tanks. Ore storage is necessary because different operations in a milling plant occur at varying rates, with some being intermittent while others are continuous in nature. In addition, some operations are batch while others are prone to frequent interruptions. For instance, mine operations and primary crushing are more prone to frequent interruptions in comparison to milling and concentration operations. It will therefore be necessary to provide storage between coarse crushers and fine grinding mills. Front-end loaders and bucket-wheel reclaimers can be used to reclaim materials from stockpiles. There are two types of stockpiles-the elongated and conical stockpiles.[5]

The total conical stockpile capacity is given by the equation:

$$C = \frac{3.14\,(tanA)\,R^3 D}{3000} \tag{2.1}$$

where
   C = capacity (tons)
   A = angle of repose for the ore material to be stockpiled
   D = density of material (kg/m^3)
   R = stockpile radius in m

This means that the capacity of the stockpile increases with the cube of the radius of the cone. An increase of 26% in radius, will double the capacity of the stockpile. For most common materials, the angle of repose is 38°. Storage allows plants to ensure a consistent feed to the processing plant since the different ores stored prior to delivery to the plant can be blended according to specifications. Slurry storage is done in conditioning tanks to allow time for reactions for froth flotation. Surge tanks are also placed along the slurry flow lines to allow for control of feed rate.[1]

2.5   Ore Feeding

The use of feeders becomes necessary when succeeding operations take place at different rates. Feeders are also required when key operations are interrupted with storage tanks. A feeder allows a uniform delivery or conveyance of dry or moist material through a gate over a short distance. Ore feeding enables a uniform stream of materials supply at a controllable rate.

A feeder typically consists of a small bin, a gate and a conveyor. Examples of feeders are aprons, belts, chains, rollers and rotary and vibrating feeders. A scalping feeder is required to avoid the "packing of the crushing chamber" that occurs due to a feed from a bin going into a jaw crusher without a preliminary fine removal leaving fines that become segregated in the feeder bin. Such segregated fines can pass through the crusher upper section and escape into the finishing zone where they block the voids between the lumps and may lead to a jamming of the crusher. Heavy-duty screens called grizzlies are therefore placed before the primary crusher to remove fines and undersize. The vibrating grizzlies scalp and feed the primary crusher in one operation. Another example of ore feeder is a surge bin for crushers. A surge bin for crushers is a device arranged to conveniently receive supplies that arrive intermittently from trucks and skip cars and to feed them steadily through gates at a controllable rate into the primary crushers of the plant. Figures 2.4 and 2.5 show an apron feeder and a brute force vibratory feeder, respectively.[6,7]

FIGURE 2.4   7" diameter Eriez apron feeder.

Source: Courtesy of Eriez (2022c)

FIGURE 2.5   Eriez brute force vibratory feeder.

Source: Courtesy of Eriez (2022d)

2.6   Ore Handling in the Laboratory

2.6.1   Ground Survey

In geochemical surveying, the baseline chemistry of soil, stream water and stream sediment is determined using samples taken by hand at a rate of about one sample every 4 km². Laboratory analyses of the samples to trace level enables a suite of maps that are useful for mineral exploration and other purposes to be produced. Two samples weighing about 1 kg each are collected at 20 and 50 cm depths. A typical Mineral Reconnaissance Program (MRP) also involves stream sediment collection, geochemical survey, soil and basal till sampling, petrological and mineralogical investigations, microchemical analysis of mineral grains and collection of geophysical data. Microchemical analysis is done to determine the bedrock from which alluvial grains originate. Figure 2.6 shows stream sediment collection during a ground survey in Northern Ireland.[8]

Samples of stream sediments taken from the stream bed, with a hand shovel, are wet-sieved to remove the finest particles in them for analysis. Water samples from the stream are also collected in bottles and some of these stream waters are tested to determine alkalinity and conductivity, while other samples are carefully filtered at the site to remove their particles before analysis. Data from samples obtained from the sites are mapped to indicate the regional trends in the changing earth's surface chemistry. Drainage sampling involves sieving stream sediment and collecting stream water and vegetation samples to obtain an indication of the trace elements present in the stream catchment area. At each site, the sediment is wet-sieved and the fine sediment passing 150 µm is allowed to settle before it is decanted into paper bags for drying and laboratory analysis. The coarse fraction residue material retained on the 150 µm sieve and that passed the 2 mm screen is then panned and observed with a hand lens for any high-density mineral present. It is then placed in a paper bag for future reference. Water samples are filtered on-site, taken into plastic containers, acidified and sent for laboratory analysis. Other stream water samples are also taken and tested for pH, conductivity and alkalinity. For the geochemistry activities, the presence of many elements and compounds is determined in the topsoil, stream water and stream sediment of the exploration environment.

FIGURE 2.6   Stream sediment collection in Northern Ireland.

Source: **Courtesy of British Geological Survey (2022)**

Borehole logging is a basic operational technique for measuring the physical and chemical properties of rocks. In drilling, rocks from various depths beneath the surface are taken out when the ground surface is penetrated and are tested to confirm the geology beneath and/or to provide samples for chemical analysis. The drill bit is attached to the longitudinal drill rod and it penetrates the ground to form boreholes from which rock samples are collected. Since the rocks produced from drilling can be identified to specific depths, the method is taken to be extremely important in determining the types and structures of the rocks underneath. It is therefore also widely used not only for mineral exploration but also in geotechnical engineering and groundwater studies. Borehole logging is the process that involves the study of the resultant samples to obtain definitive boundaries between different rock types and thus develop the picture of the geology below the surface.

Assaying is the process of analysing a rock sample in a laboratory to determine its precise chemical composition and it is particularly important in investigating the metal contents of potential ore bodies. The dimension of the borehole usually ranges between 10 and 30 cm in diameter and can be as long as tens of meters to hundreds of meters. Reverse circulation (RC) and diamond drilling are the most common types of drilling methods for mineral exploration. Diamond drilling is a very expensive method of drilling; however, it has the advantage that it can be used to determine the nature and angle of the ore body at its margins where it is in contact with the surrounding rock types. Therefore, despite its high costs, it may be required because it can provide vital details that RC drilling cannot.

Due to the high costs of drilling, it is usually the last step to be carried out in the mineral exploration process when the preliminary work has yielded results that are sufficiently successful. Drilling may then be finally used to confirm any mineralization physically and also to develop the geological picture. A decision on mine development can be made after drilling a sufficient number of holes to confirm the presence or otherwise of an ore body of a certain size and quality.[9]

2.7   Laboratory Sample Preparation

When the sample from the mine or field exploration is received in the Laboratory, it must be prepared in such a way that the work sample to be used for analysis and processing is actually representative of the bulk received. Jones riffler and coning and quartering are used in the laboratory to progressively reduce the mass of the bulk ore to work samples while avoiding a systematic bias that can affect accuracy during analyses. The technique involves preparing the sample so that it takes a conical shape and then flattening it out.

2.7.1   Jones Riffle Splitter

Before a bulk ore material is split to obtain a sample that truly represents the bulk, the bulk sample is repeatedly divided and recombined severally. The material should be able to flow and its particles' sizes should be less than one-third of the splitter chute's width. The material is poured into the hopper, by carefully pouring from one end to the other and ensuring an even volumetric flow. Upon completion of pouring, the hopper's bottom gate is opened fast using the hand lever. After the splitting, the collecting pans on each side of the splitter will contain about 50% of the material delivered within 2.5–3% error margin. The precision Jones riffle splitter is better than the traditional type in terms of its gate design that eliminate the error arising from pouring. Figures 2.7, 2.8 and 2.9 show the Eriez Jones riffle splitters.[10–12]

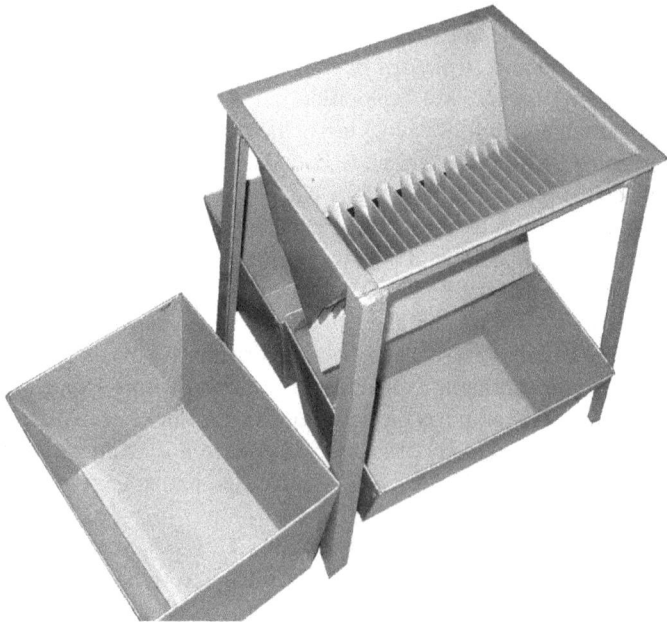

FIGURE 2.7   Figure Eriez Jones riffle splitter.

Source: Courtesy of Eriez (2022e)

FIGURE 2.8   Eriez Jones riffle splitter model SP 173.

Source: **Courtesy of Eriez (2022f)**

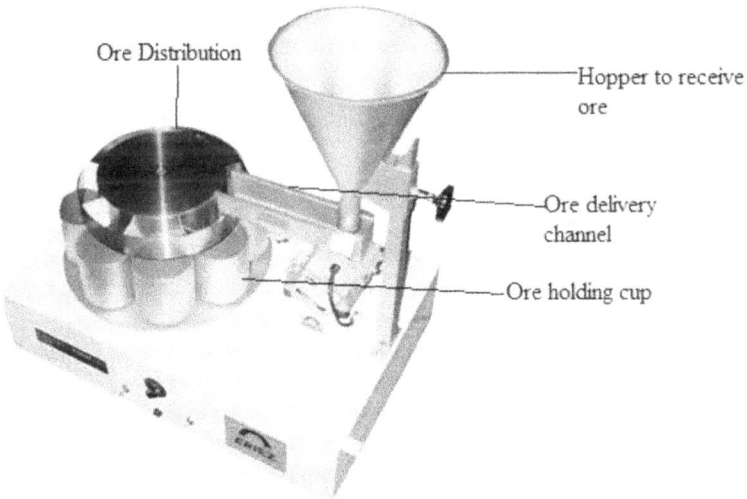

FIGURE 2.9   Eriez cascade splitter.

Source: **Courtesy of Eriez (2022g)**

2.7.2   Coning and Quartering

The procedure involves pouring about 50 kg bulk sample on a clean steel plate or a plastic sheet. The sample is spread out and mixed thoroughly into a conical heap and split into two equal halves with a steel bar. The splitting is repeated in a perpendicular direction to split the sample into four. The two opposite quarters are taken for a repeat of the procedure, while the other two are discarded. The procedure is repeated with sample collection taking place in the alternate diagonal until the desired weight of the work sample is obtained. The discarded portion is collected for storage.

2.7.3   Application of Gy Sampling Theory

The founder of the Theory of Sampling (TOS), Pierre Gy (1924–2015) developed the Theory of Sampling from 1950 to 1975 single-handedly and spent the 25 years that followed applying it in some key industrial sectors such as mining, minerals, cement and metals processing. Gy was able to identify at least eight sampling errors that indicate all that can go wrong in sampling, sub-sampling (sample mass reduction), sample preparation and sample presentation as a result of heterogeneity and/or inferior sampling equipment design and performance. Since samples in mineral processing are heterogeneous in size and mineral type, it is reasonable to apply Gy theory to minimize experimental errors that may relate to improper sampling. It requires several millions of dollars to construct a mineral processing plant and a large outlay of expenses to run it. However, the success of the plant depends on the optimization of its operation which in turn depends on the results obtained from small work samples obtained from various ore sources. It is therefore very important that care be taken to ensure that the small fraction that constitutes a sample should be a true representative of the bulk material as much as possible. In view of this, much effort in statistical analysis and sampling theory is applied in any sampling process to quantify the procedures and precautions required.[13, 14]

The probability of a sample truly representing the bulk has been found to increase when incremental samples are taken while collecting from a stream of material flowing during processing such as from a conveyor belt for solids and off pipes for liquids or slurries. In order to increase the probability of a work sample adequately representing its bulk, several methods have been proposed. One such method that involves the determination of the work sample to be taken to ensure the accuracy of the analysis to be carried out is the Gy equation. Gy introduced a model based on equiprobable sample spaces and proposed that if:

d_{MAX} = dimension of the largest particle
M_{MIN} = minimum mass of sample required
σ^2 = variance of allowable sampling error in an assay (in the case of a normal distribution, this equals the standard deviation), then:

$$M_{MIN} = \frac{Kd_{MAX}^3}{\sigma^2} \tag{2.2}$$

where K is called the sampling constant (kg/m).

In mineralogical sampling, the dimension of the largest particle piece in the sample (d_{MAX}) can be taken as the screen aperture through which 90–95% of the material passes. As $\pm 2\sigma$ represents the probability of events when 95 out of 100 assays would be within the true assay value, 2σ is the acceptable probability value of the sample. The sampling constant K is considered to be a function of the material characteristics and is expressed by:

$$K = P_S P_D P_L \cdot m \qquad (2.3)$$

where P_S = particle shape factor (usually taken as 0.5 for spherical particles, 0.2 for gold ores)

P_D = particle distribution factor (usually in the range 0.20–0.75 with higher values for narrower size distributions, usually taken as 0.25 and 0.50 when the material is closely sized

P_L = liberation factor (0 for homogeneous (unliberated) materials, 1 for heterogeneous (liberated) and Table 2.1 presents the liberation factor for intermediate materials.

m = mineralogical factor

The mineralogical factor, m, has been defined as:

$$m = \frac{\left[(1-\alpha)\right]}{\alpha}\left[(1-\alpha)\rho_m + \alpha\rho_G\right] \qquad (2.4)$$

where α is the fractional average mineral content and ρ_m and ρ_G the specific gravity of the mineral and the gangue, respectively.

The liberation factor, PL, is related to the top size, dMAX and to the liberation size, dL of the mineral in the sample space. In practice, PL is rarely less than 0.1 and if the liberation size is unknown, then it is safe to take PL as 1.0.

When a large amount of sample has been collected, it has to be split by a suitable method such as riffling. At each stage of subdivision, samples have to be collected, assayed and statistical errors determined. In such cases, the statistical error for the total sample will be the sum of the statistical errors during sampling (σ_S) and the statistical error in an assay (σ_A) so that the total variance (σ_T^2) will be:

$$\sigma_T^{\,2} = \sigma_S^{\,2} + \sigma_A^{\,2} \qquad (2.5)$$

TABLE 2.1

Liberation F for Intermediate Materials

d_{MAX}/Dl	<1	1–4	4–10	10–40	40–100	100–400	>400
PL	1.0	0.8	0.4	0.2	0.1	0.05	0.02

When the sample is almost an infinite lot and where the proportion of mineral particles has been mixed with gangue and the particles are large enough to be counted, it may be easier to adopt the following procedure to determine σ_p. Let

P_M = proportion of mineral particles
P_G = proportion of gangue particles
N = number of particles

Then the standard deviation of the proportion of mineral particles in the sample, σ_p, will be:

$$\sigma_p = \sqrt{\frac{P_M P_G}{N}} \tag{2.6}$$

The standard deviation on a mass basis (σ_M) can be written in terms of the per cent mineral in the whole sample provided the densities of the mineral constituents are known. Thus if ρ_M and ρ_G are the densities of the mineral and gangue, respectively then the mass per cent of mineral in the entire sample, consisting of mineral and gangue will be:

$$A_M = \frac{100 P_M \rho_M}{P_M \rho_M + P_G \rho_{GM}} \tag{2.7}$$

If it is assumed that the particles of mineral value and gangue have the same shape and size the standard deviation of the entire sample will be given by Gupta and Yan (2016)[14]:

$$\sigma_T = \frac{dA_M}{dP} . \sigma_P$$

or

$$\sigma_T = \left(\frac{(100 - A_M) P_M + A_M P_G}{100 \sqrt{\rho_M \rho_G}} \right) . \sqrt{\frac{A_M (100 - A_M)}{N}} \tag{2.8}$$

Example 2.1
   Copper ore samples with a copper content of about 10% are required to feed a copper processing plant. The confidence level for the experimental determination was required to be 0.15% Cu at 2σ standard deviations. The Cu-bearing mineral, chalcocite in the ore was found to have a liberation size of 75 um, while the top size of the ore from the sampler was 2.6 cm. Given the density of chalcocite and gangue as 5600 and 2650 kg/m³, respectively, determine the minimum mass of sample required to represent the ore.

Solution

From the data, dMAx = 2.6 cm

Since the confidence interval required is equal to ± 0.15% of a 10% assay,

$2\sigma = 0.15/10$ or $\sigma = 0.015/2 = 0.00750$

dL = 75 um = 0.075 mm

Again from the data dMAx/dL = 26/0.075 = 346.67; hence from Table 2.1,
   PL = 0.05

Let the mass of the ore = 100 kg

The mass of the Cu in the ore = 0.1×100 kg = 10 kg

The mass of chalcocite = ?

Molar mass of chalcocite = 2X64 + 32 = 128 + 32 = 160 g

Mass of chalcocite = $\sigma \dfrac{160}{2 \times 64} \times 10 = 12.5\,g$

The ore thus contains 12.5 wt% Cu_2S and thus $\alpha = 0.125$

The mineralogical composition factor, m, can now be calculated from the equation:

$$m = \frac{\left[(1-\alpha)\right]}{\alpha}\left[(1-\alpha)\rho_m + \alpha\rho_G\right]$$

$$m = \frac{\left[(1-0.125)\right]}{0.125}\left[(1-0.125)\times 5600 + 0.125 \times 2650\right]$$

$$= 7 \times \left[(4900+331.5)\right]$$

$$= 7 \times 5231.5 = 36618.75\ kg/m^3$$

$$K = 0.5 \times 0.25 \times 0.05 \times 36618.75 = 228.87\ kg/m^3$$

$$\text{But } M_{MIN} = \frac{Kd_{MAX}^{3}}{\sigma^2} = \frac{228.87 \times 0.026^3}{0.00750^2} = 71.51\,kg$$

Thus the minimum sample size to be taken to ensure that the analysis results to be obtained satisfy the specification indicated should then be 71.51 kg.

2.8 Determination of Moisture Content

It is important in mineral processing to determine the dry weight of the ore sample. The accurate measuring of moisture in dry bulk materials is one of the key steps to ensure that the final product from the raw material will meet specific quality requirements. The methods of moisture content determination can be categorized as laboratory methods, in which a material sample is analyzed in a laboratory away from the process line, and inline methods, in which data from a remote moisture sensor (or another measuring device) mounted on a material storage vessel or process equipment in the process line is used to calculate the material's moisture content.

In the direct laboratory methods, the moisture in the dry bulk solid material is subjected to a reaction or measurement. However, the methods take a long time and the accuracy depends on taking a truly representative sample of the bulk solid. These methods include the analytical method known as Karl-Fischer titration and three methods based on gravimetric analysis, that is, the dry chamber, microwave drying and infrared drying methods. The work sample for moisture content determination is taken by grab sampling. This involves taking small quantities of the material from different spots in the bulk and mixing them thoroughly to form the final work sample.[15]

2.8.1   The Karl-Fischer Titration Method

The Karl-Fischer titration method is used to analyse small samples of solid materials such as minerals, paint pigments, plastic granules and cloth fibres that contain low moisture. Karl-Fischer solution consists of a mixture of a base, iodine and sulphur dioxide dissolved in a solvent such as alcohol. The method is based on a titration reaction called Karl-Fischer titration chemical reaction with the simplified equation:

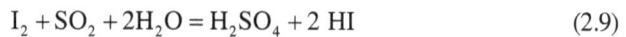

$$I_2 + SO_2 + 2H_2O = H_2SO_4 + 2\,HI \tag{2.9}$$

For moisture content analysis, the water in a sample is removed from the solids with methanol and analyzed with the Karl-Fischer solution (I_2 + SO2 + H_2O) absorbed by the sample. The reactant compound, I2, which is red-brown in color is converted to hydrogen iodine compound, HI, which is transparent in color. The moisture content of the reaction's final titration product, I_2, is measured by photometric or electrometric means. The sample's water content is taken as the quantity of Karl-Fischer solution absorbed.

2.8.2   Drying Oven Method

A one-gram representative sample of the material is taken with an electronic balance. The sample is dried at 110°C ± 5°C for 24 hours in the dry chamber and then re-weighed after cooling to about 30°C. The sample is dried for another 24 hours and weighed again. If the weight loss is less than 1 per cent, the sample is taken as dry, otherwise, the drying continues until the weight loss becomes less than 1 per cent. There are other modifications to this procedure.

2.8.3   Microwave Oven Drying

The microwave drying method is another form of gravimetric analysis that involves drying and weighing the sample typically one handful or a cupful of material, depending on the material. In this method, microwaves dry the sample in a very short time typically a few minutes. Since the sample is heated to very high temperatures of between 140°C and 750°C, the method is suitable only for materials that can withstand high heat. The method cannot be used for materials such as sugar, gypsum and aluminium hydrate that can be destroyed at these temperatures or whose physical composition can be changed by high heat. The weight loss after heating gives the moisture content.[16]

2.8.4 Inline Methods

Inline methods are based on technologies such as capacitive, infrared, microwave, neutron and gamma rays as well as conductivity. It requires the conversion of data about the behavior of moisture or other physical structures in the material to indirectly calculate the material's moisture content and is thus called the indirect method. One or more remote moisture sensor (or another measuring device) is required to collect the data, and a controller, which can be of various types, uses the data to calculate the material's moisture content. Because the moisture measurement is instant, these methods can be used to continuously monitor the moisture content of the material in a process line. For the determination of moisture content in coals, there are standard specifications as reported by Francis and Peters (1980).[17]

2.8.5 Determination of Loss on Ignition

Two 50 ml silica crucibles are placed in an electric furnace and heated to 950°C and retained for at least 30 min at the set temperature. The heated crucibles are then removed and placed in the desiccator to cool to ambient temperature and then weighed to the nearest 1 mg (M_a). 1 g of the dry sample is to be weighed into each of the crucibles (M_b) and heated in the furnace at 950°C for at least 60 min. The hot crucibles carrying the residue on ignition are to be placed in the desiccators for cooling. After cooling to ambient temperature, the crucibles with the dry residue are weighed to the nearest 1 mg (M_c). The crucibles are to be quickly weighed immediately after removal from the desiccators. The weight of the residue on ignition and therefore the loss on ignition is taken as constant if the weight obtained after a further 30 min of ignition at 950°C in the preheated furnace ($M_c - M_a$) differs by not more than 0.5% of the previous value. The loss on ignition (W_v) is determined using the equation.[18]

$$W_v = \frac{M_b - M_c}{M_b - M_a} \qquad (2.10)$$

2.9 Ore Roasting

A 50 g mass of the ore is to be placed in a silica dish with dimensions 145 × 85 × 20 mm to occupy three quarters of the silica dish. The sample is then placed in a cold furnace and set at 350°C to commence heating. The furnace's door is to be set 25 mm wide open throughout the roasting period to ensure a free supply of oxygen. The sample is to be turned after every 15 min interval for efficient mixing and the temperature is raised by 100°C every hour until the roasting temperature is attained. This is to be done in four steps that are about 25°C every 15 min. The roasting temperature is typically taken as 750°C.[19]

References

1. Wills, B.A., and Napier-Munn, T.J. (2006): *Mineral Processing Technology*, 7th Edition. Amsterdam: Elsevier Science and Technology Books.
2. 911 Metallurgists (www.911metallurgist.com/underground-ore-loading-chutes/, Accessed 27th May, 2020).
3. Eriez (2022a): *Magnetic Belt Conveyor* (www.eriez.com/Images/Product-Images/ Material-Handling-Equipment/Magnetic-Belt-Conveyors/MagneticBeltConveyor.JPG, Accessed 17th April, 2022).
4. Eriez (2022b): *Industry Ore Conveying Systems Example* (www.eriez.co.za/ZA/EN/ Eriez/Industries/Mining-Minerals-Processing.htm, Accessed 30th May, 2022).
5. 911 Metallurgist (www.911metallurgist.com/blog/laboratory-methods-of-sample-preparation, Accessed 3rd September, 2020).
6. Eriez (2022c): *7" Diameter Eriez Apron Feeder* (www.phxequip.com/Multimedia/ images/equipment/optimized/7-x-17-sanitary-ss-portable-vibrating-sifter-10416.jpg, Accessed 17th April, 2022).
7. Eriez (2022d): *Brute Force Mechanical Vibratory Feeder* (www.eriez.com/Images/ Product-Images/Vibratory-Feeders-Conveyors/Heavy-Duty-Feeders/BruteForce064. jpg?Medium, Accessed 29th May, 2022).
8. British Geological Survey (BGS) (2022): *Exploration Techniques* (https://www2.bgs. ac.uk/mineralsuk/exploration/techniques.html, Accessed 11th May, 2022).
9. Geological Survey of Ireland (GSI) (www.gsi.ie/en-ie/programmes-and-projects/tellus/ activities/ground-survey/Pages/default.aspx, Accessed 27th May, 2020a).
10. Eriez (2022e): *Eriez Jones Riffle Splitter* (www.eriezlabequipment.com/lab-equipment/splitting-and-dividing/jones-rifflers/, Accessed 14th May, 2022).
11. Eriez (2022f): *Eriez Jones Riffle Splitter Model SP 173* (www.globalgilson.com/ jones-riffle-splitters, Accessed 21st May, 2022).
12. Eriez (2022g): *Cascade Splitter* (www.eriez.com/Images/Product-Images/Size-Reduction-Equipment/CascadeSplitters.jpg?Medium, Accessed 25th April, 2022).
13. Esbensen, K.H. (2018): *Pierre Gy (1924-2015): The Key Concept of Sampling Errors* (www.spectroscopyeurope.com-sampling-pierre-gy, Accessed 4th September, 2020).
14. Gupta, A., and Yan, D. (2016): *Mineral Processing Design and Operations*, 2nd Edition. Amsterdam: Elsevier.
15. Sepor (www.sepor.com/product/rotary-sample-splitters/, Accessed 27th May, 2020).
16. Spitzlei, M. (2002): *Choosing a Method for Measuring Your Material's Moisture Content* (file://C:\Documentos\Download\ArtigosWEB\Choosing%20a%20method% 20for%20 . . . , Accessed 3rd September, 2020).
17. Francis, W., and Peters, M.C. (1980): *Fuels and Fuel Technology*, 2nd Edition. New York: Pergamon Press.
18. Heiri, O., Lotter, A., and Lemcke, G. (2001): Loss on Ignition as a Method for Estimating Organic and Carbonate Content. *Journal of Paleolimnology*, 25, 101–110.
19. Gilchrist, J.D. (1989): *Extraction Metallurgy*, 3rd Edition. Oxford: Pergamon Press, p. 145.

3 Ore Crushing

3.1 Introduction

Ore comminution is a two-stage mechanical operation and it entails crushing as the first stage and grinding as the second and the main objective is to release or liberate particles of the mineral value(s) from the particles of the associated undesirable gangue minerals.

The important issues relating to ore crushing are:

1. It is generally carried out in a dry state and takes place in two or three stages called primary, secondary and tertiary crushing and uses heavy-duty machines to reduce ore lumps as large as 1.5 m or more across in size to lumps of 10–20 cm in size.
2. Due to delays relating to occasional insufficient supply of run-of-mine ore and mechanical interruptions arising from the possible breakdown, primary crushers are usually designed to operate at 75% of the available operating time.
3. In secondary and tertiary crushing, the primary crusher's product is treated to produce a crushed product's final discharge with a size ranging between 0.5 and 2 cm.
4. For ores that are tough and slippery, coarse grinding in the rod mill may replace the tertiary crushing stage of primary crushing.
5. For ores that are extra hard and also to avoid the generation of too many fines, the secondary crushing stage may involve more than two reduction stages.
6. A washing stage is included in the crushing plant to wash slimes and clays from sticky ores.
7. The secondary crusher is usually preceded with vibrating screens to prevent the entry of undersized material or to scalp the feed into it. The purpose is to increase the crushing capacity by not allowing undersize particles in and to avoid fines getting stuck between the lumps and thus choking the crushers.
8. For a mineral processing plant, a typical crushing flowsheet will have from one-to-three stages of crushing.[1]

3.2 Primary Crushers

The two main types of crushers used for metalliferrous ores, that is, ores containing or yielding metals, are jaw and gyratory crushers, while impact crusher also has limited use as a primary crusher. The typical operating parameters of a crusher are feed rate, feed size and closed side setting, that is, gape setting. The following are important points to note on the primary crusher:

DOI: 10.1201/9781003323433-3

- The primary crusher's purpose is to reduce the ROM ore to a size suitable to be fed into the secondary crusher or the SAG mill grinding circuit
- A primary crusher can achieve a reduction ratio of up to about 8:1 or more
- Feed depends on the upstream process output and is typically Run-of-Mine (ROM) of up to 1.5 m in size
- Product:
  - < 300 mm fraction (for transport) to -200 mm (for feeding the SAG mill) and is determined by the gape setting
- Feed Rate depends on the feeder and is typically 160–13,000 tons per hour (tph)
- Typical rules for primary crusher selection:
  - Rule 1: Due to lower costs, a jaw crusher is to be preferably used always, if possible
  - Rule 2: For low-capacity applications, the oversize is to be preferably crushed using a jaw crusher and hydraulic hammer
  - Rule 3: For high-capacity applications, the use of a jaw crusher with big intake openings is recommended
  - Rule 4: For very high-capacity applications, a gyratory crusher is the recommended choice
- The reduction ratio in crushing indicates the working ability of a crusher.

The following should also be noted:

For ore feed that is soft and is not abrasive, the horizontal shaft impactor (HSI) crusher is preferred, particularly for moderate crushing capacity.

Jaw crushers require the least capital cost and are thus recommended for use whenever possible

For low-capacity operations, it is better to use a jaw crusher and hydraulic hammer for oversize

For high capacity operations between 800 and 1500 tph, jaw crusher with big intake opening is recommended

For very high capacities operations such as those higher than 1200 tph, a gyratory crusher is preferred

For rock and gravel crushing, the final crushing stage is very important as it determines the final size and shape of the product. It is therefore recommended to use either a cone crusher, a vertical shaft impact (VSI) crusher or High pressure grinding rolls (HPGRs). The maximum feed size is determined by the crusher chamber, while the product shape depends on the setting/speed.[2]

3.2.1 Jaw Crushers

The distinguishing feature of the jaw crusher is that it has two jaws which are placed with an acute angle between them and the two open and close as animal jaws do. The jaws are such that one is fixed while the other is hinged, swinging relative to the immobile jaw. The jaws open and close 250–400 times per minute. When an ore lump is fed into the jaw crusher, the lump is alternately nipped and released until it

FIGURE 3.1 Eriez jaw crusher.

Source: Courtesy of Eriez (2022a)

exits the jaws. Jaw crushers are classified based on the way the swinging jaw is pivoted. In the Blake Jaw crusher, the swinging jaw is pivoted at the top and thus it has an ore receiving area that is fixed and an exit or discharge opening that is variable. For the Dodge Toggle Jaw crusher, on the other hand, the pivoting is at the bottom making it to have a receiving area that varies and a delivery opening that is fixed. By its construction, the dodge crusher chokes easily and is therefore restricted to laboratory use to produce closely sized products. The Universal Jaw Crusher is however hinged at an intermediate point and thus has the receiving and discharging areas that vary. Figure 3.1 shows the Eriez jaw crusher designed to crush materials of moderate hardness such as coal, limestone, bricks, ores and rubbles.[3]

3.2.2 Blake Jaw Crusher

There are two types of Blake jaw crushers, the single and the double toggle jaw crusher. In the double toggle Blake jaw crusher, the Pitman moves up and down under the influence of the eccentric to cause the oscillating motion of the swinging jaw, while the upward pushing of the toggle plate causes the Pitman to move sideways. The front toggle plate receives the motion generated and causes the moving jaw to close on the immobile jaw. Figure 3.2 shows a jaw crusher's crushing action.[5]

The swing jaw opens when the pitman undergoes a downward movement. In the single toggle Blake jaw crusher, the eccentric shaft has a swing jaw that moves towards the fixed jaw suspended on it. It also moves in the vertical direction acted upon by the toggle plate as the eccentric flywheel rotates. The motion of the elliptic jaw crusher aids to push the ore lump into the chamber where crushing occurs. It also improves the single toggle jaw crusher's capacity above that of the double toggle crusher. The main features of the Blake jaw crushers are:

1. It moves the shortest distance at the entry and the longest distance at the exit since its pivot is at the top.

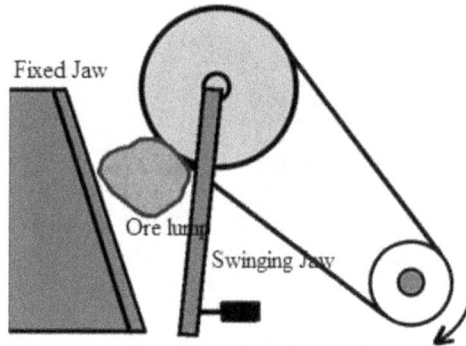

FIGURE 3.2 Jaw crusher crushing action.

Source: **Courtesy of Wikimedia (2022), adapted**

2. At the bottom of the Pitman's cycle, the displacement of the swinging jaw horizontally attains its maximum and this displacement reduces in a steady manner as the angle between the back toggle plate and the Pitman reduces through the rising half of the cycle.
3. At the onset of the cycle when the angle between the toggles is lowest, the least crushing force is exerted and the exerted force becomes highest at the top when full power is obtained over a decreased movement of the jaw.

Jaw crushers are rated on the basis of the receiving areas, which are the product of the width of the jaws and the gape. The gape is the distance at the feed opening between the jaws. For instance, a 2130 × 1680 mm jaw crusher specification means that the jaw crusher has a width of 2130 mm and a gape of 1680 mm. A rock piece that falls into the crusher's mouth, is nipped by the jaws moving relative to each other at a rate of 100–350 rev/min determined by the crusher's size. The piece is released, allowed to fall and arrested again in a repeating pattern. The swing jaw closes on the piece, firstly quickly and after the closing, it proceeds at a slower rate with increasing force towards the stroke's end. For a given crusher, the gape and the width are set values while the setting is variable to obtain the product size desired.[1, 2]

Figure 3.3 presents a well-labelled jaw crusher.[6]

The dimensions of the jaw crusher are as shown in Figure 3.4.[6]

The following definitions should be noted:

- Gape: The distance between the jaws at the feed-receiving opening
- Closed side set (CSS): it is the minimum discharge aperture size and the minimum opening between the jaws during the crushing cycle
- Open side set (OSS): The maximum discharge aperture size
- Throw: it is the difference between OSS and CSS and it is the stroke of the swing jaw

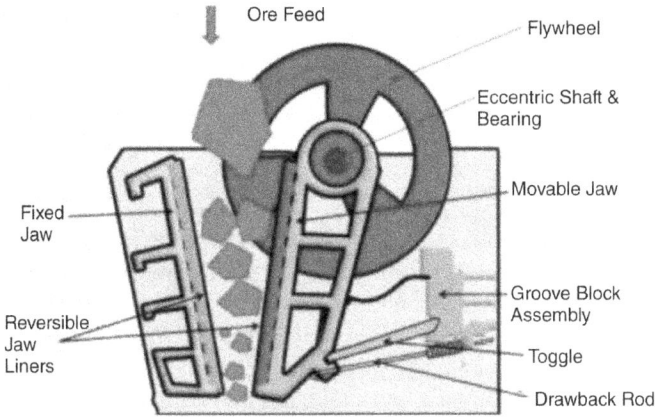

FIGURE 3.3   A labelled jaw crusher.

Source: **Courtesy of 911 Metallurgist (2022a)**

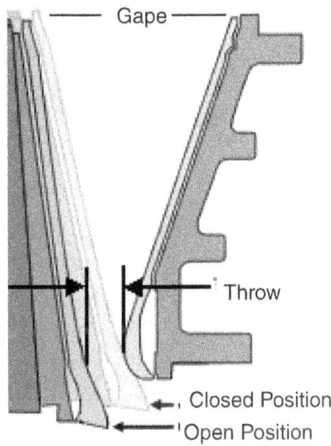

FIGURE 3.4   Dimensions in a jaw crusher.

Source: **Courtesy of 911 Metallurgist (2022b)**

A jaw crusher's typical fundamental specifications are:

- Vertical height must be 2 × Gape size where gape is given in meters
- Width of jaw should be > 1.3 × Gape size but should be < 3.0 × Gape size
- Throw = 0.0502 × (Gape size) × 0.85

Jaw crusher sizes and power ratings are specified as follows:

- Size is specified as gape size × width
- The largest capacity jaw crusher is the one with the size specification 1600 × 2514 mm having motor ratings of 250–300 kW

- Metso crushers (C200 series) are of 1600 × 2514 mm size specification but with a much higher power rating of 400 kW
- The largest particle that the opening of the jaw crusher can take in is estimated by: 0.9 × gape

3.2.3 Jaw Crushers Feeding

In the mineral processing plants, run-of-mine feeds are often provided on an intermittent basis by trucks, shovels or front-end loaders. However, it is desired to provide feed into the jaw crushers in a continuous mode instead of intermittent one as the ores arrive and thus it becomes necessary to use feeders. When the rate of ore feeding surpasses the rate of crushed ore production, the condition called choke feeding is obtained. In choke feeding the crushing occurs by particle breakage taking place among the particles interacting and between the crusher's plates and the particles. The output of choke feeding is fine which is desired for mineral value particle liberation. Jaw crushers' advantages over gyratory crushers include lower cost of installation and their ability to resist high abrasion without much maintenance. However, it has the disadvantages of being useable for primary crushing only and a comparably low maximum capacity of 1,000 metric tons per hour (mtph) in comparison to a gyratory crusher of the same gape.

3.3 Gyratory Crusher

The gyratory crusher comprises essentially of a long spindle equipped with the head, a hard conical grinding component seated in an eccentric sleeve. The spindle is suspended from a "spider" and due to the gyratory action of the eccentric, it rotates between 85 and 150 rev/min moving in the fixed crushing chamber along a conical path. The gyratory crusher is designed such that there are effectively two sets of jaws opening and shutting like a jaw crusher at any cross-section of the crusher. Heavy steel castings or welded steel plates are used for the construction of the outer shell of the crusher. The spider holds a suspended bearing which mainly carries the spindle. The gyratory crusher does its crushing on a full cycle unlike the jaw crusher and thus it has a higher crushing capacity than a jaw crusher having the same gape. They are thus preferred in plants with large production rates such as those with a crushing rate of 900 tph.

Gyratory crushers range in capacity size of up to 1830 mm gape and can crush ores with top size as large as 1370 mm at a crushing rate of 5000 tph with a 200 mm set. When a rock lump feed descends into the crusher, it is squeezed by the mantle against the concave surface and as the mantle departs, the rock descends lower in the chamber and becomes subjected to further crushing in the next cycle. The open side setting is the maximum discharge exit for the ore feed and therefore its setting determines the top size of the final product.

A flared shell or frame supports is a conical element to create a chamber with a narrow cross-section at the bottom and a wide one at the top. The centre element is made to advance and retreat in relation to the shell because of its gyration about its fulcrum point. The rock enters at the top and gets fractured as it makes its way through the crusher chamber. The gyratory crusher has a reduction ratio that ranges between 3:1 and 10:1 and an output of 300–9100 tph. The gyratory crusher design includes the following features:

- It is rare to install large units underground
- Its feed is provided by tip wagons, trucks, conveyor belts and side dump cars
- A hopper receives its feed and delivers it into a grizzly screen for oversize particles removal
- Gyratory crushings are carried out in an open circuit instead of a closed circuit
- The maximum ore feed size is restricted to sizes between 1 and 1.5 m
- The input rock lump size is crushed to sizes between 10 and 15 cm
- Typical reduction ratios are:
  - Primary crusher: 3:1–10:1
  - Secondary crusher: 6:1–8:1

The spindle carries a steel forging crusher head shielded from damage by a manganese steel mantle, The spindle is fitted into the crusher head with nuts on pitched threads so that they get self-tightened during crushing. The crusher's mantle has soft materials such as zinc, white metal or plastic cement as backing to obtain even seating against the steel bowl. The vertical profile is made bell-shaped to aid in the crushing of ores with a tendency to choke the crusher. A jaw crusher is preferred to a gyratory crusher when the crushing capacity is not as important as the gape selection. For example, when it is desired to crush an ore with a particular top size, a gyratory crusher with the required gape size will have three times the capacity of a jaw crusher with the same gape. In that case, if a high capacity is required then the gyratory crusher will be selected but it will be more economical to use a jaw crusher to satisfy the gape requirement. In plant design, the following Taggart relationship holds:

If tph < 161.7 × (gape size in m)^2, then it is recommended to use a jaw crusher
   and if otherwise, a gyratory crusher

A gyratory crusher is uniquely defined by the diameter of the mantle and the dimensions of the feed opening.

The advantages of using a gyratory crusher are:

- It is designed for direct delivery of ores from trucks with up to 300 tons capacity
- It has a high capacity rating
- The maintenance cost per ton of processed ore is considered to be the lowest
- It has the highest use availability
- It can be used to crush ores with strength in compression as high as 600 MPa (90,000 psi)

However, it has the highest installed capital cost of all primary crushers. It should be noted that in jaw and gyratory crushers, the reduction occurs progressively through

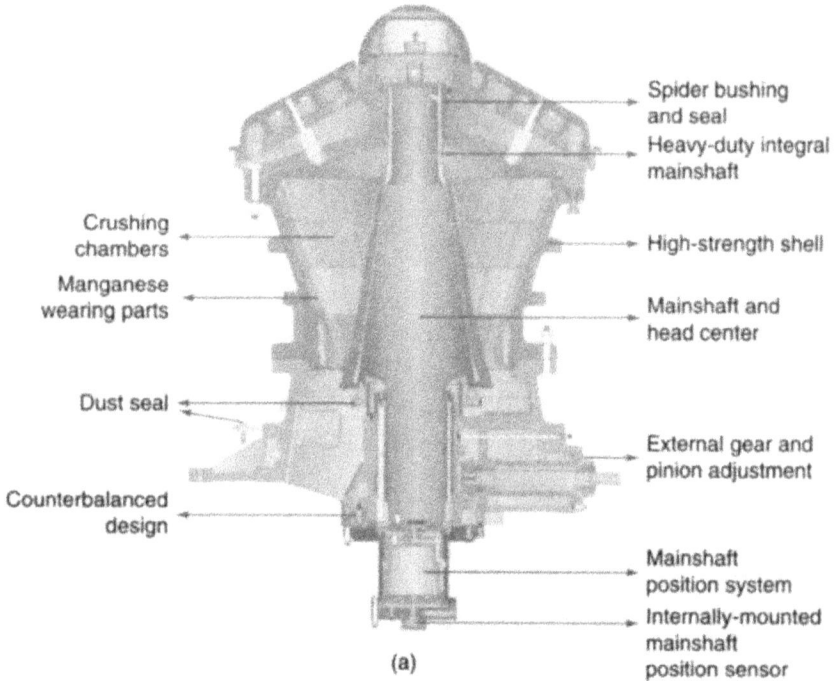

Spider bushing
and seal

Heavy-duty integral
mainshaft

High-strength shell

Mainshaft and
head center

External gear and
pinion adjustment

Mainshaft
position system

Internally-mounted
mainshaft
position sensor

Crushing
chambers

Manganese
wearing parts

Dust seal

Counterbalanced
design

(a)

FIGURE 3.5   Gyratory crusher.[8]

Source: **Courtesy of 911 Metallurgist (2022c)**

the exertion of repeated crushing force as the rock lump descends towards the discharge point UAF.

3.4   Secondary Crushers

Secondary crushers such as cone crushers, hammer mills and crushing rolls receive the primary crushers' dry output with sizes less than 15 cm in diameter. They are used to reduce the primary crushers' outputs to sizes suitable for grinding. Tertiary crushers have the same design as secondary crushers but the set is closed.

3.4.1   Roll Crusher

Roll crushers or crusher rolls have applications in crushing ores such as soft iron ores, chalk, limestone, gypsum and phosphate that may be sticky, friable, frozen and less abrasive. They are preferred to gyratory and jaw crushers that tend to experience choking near the discharge point when treating friable rocks with a high percentage of large-size lumps. Roll crushers have two standard spring rolls comprising of two cylinders placed horizontally and revolving towards one another. Unlike in gyratory crushers and jaw crushers, roll crushers' crushing occurs in one single pressure. Rolls with smooth surfaces are applied in fine crushing, while corrugated rolls and

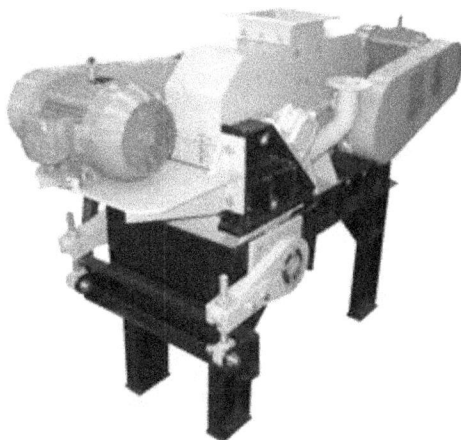

FIGURE 3.6   Eriez MACSALAB rolls crusher.

Source: Courtesy of Eriez (2022b)

those with stub teeth are used in coarse crushing. Figure 3.6 shows the Eriez roll crusher which breaks rocks by nipping and has the capability to reduce coal, ores and hard rock from 20 mm top size to fine sizes with not many fines produced.[1,8]

3.4.2   Cone Crusher

The design and development of the cone crusher were by Symons in 1920. It is similar to a gyratory crusher except that instead of the spindle being suspended it is supported at the bottom of the gyrating cone. The head-to-depth ratio is larger than for gyratory crushers and the cone angles are flatter. In addition, concaves and mantle slopes are parallel and finer product particles result from the flatter cone angles that increase lumps' residence time. The passage of rocks that cannot be broken occurs because the shell is held by springs. Since it receives smaller lumps and thus needs a small gape, the crushing shell or bowl flares out and thus allows for the swelling of the crushed ore as it provides an increasing cross-section as the ore descends. A replaceable mantle with zinc, cement or epoxy resin backing protects the crushing head. The diameter of the cone lining which typically ranges from 559 to 3100 mm is used to rate the crushing head. Figure 3.7 presents the image of a cone crusher crushing head showing its components.[10]

Figure 3.8 shows a cone crusher machine.[11]

3.4.3   Impact Crusher

The crushing in the impact crusher is by impact, not compression as for the crushers mentioned earlier, and it is by sharp blows exerted in rapid succession to rocks falling freely through the chamber. The impact creates stresses in the particles large enough to cause them to shatter. The lumps broken by compressive pressure retain residual stresses that can cause cracking later, while the impact on the other hand causes the lump to fracture

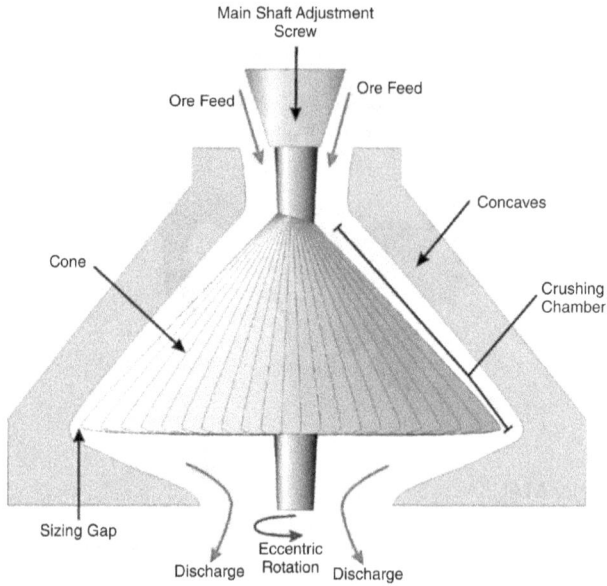

FIGURE 3.7   A cone crusher showing the main components.

Source: **Courtesy of British Geological Survey (2022), adapted**

FIGURE 3.8   A cone crusher.

Source: **Courtesy of 911 Metallurgist (2022d)**

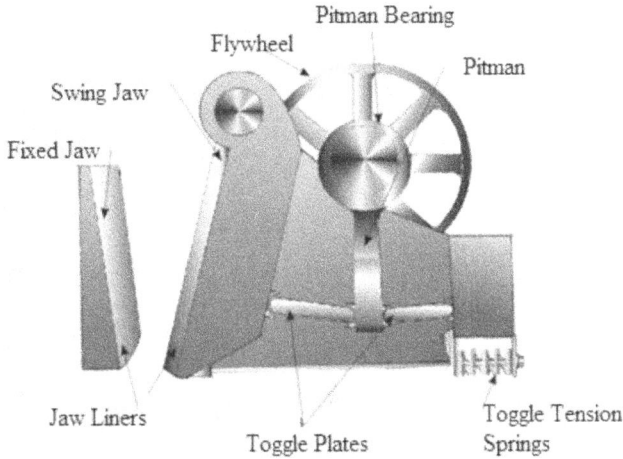

FIGURE 3.9   Impact crusher.

Source: Courtesy of 911 Metallurgist (2022a, 2022e)

immediately without residual stresses. The internal stress-free states of the crushed products make the impact crusher preferable for breaking stones for building, brick making and roadmaking. Impact crushers are also preferred in quarrying because they produce products with higher elongations as a result of the high reduction ratios achievable.[1]

The advantages of using impact crushers are:

* It enables a large size reduction from very large, say 1000 to 75 mm
* The reduction ratio achievable is high for the level of investment required
* For ores, impact crushers produce a high proportion of fines
* The tonnage output from the impact crusher can be as high as 2500 mtph

The disadvantages are:

* Since more fine products are obtained, power consumption is comparably higher
* A feeder is required

Figure 3.9 shows an impact crusher:[6,12]

3.4.4   Hammer Mills

It consists of breaker plates and hammers produced from abrasion-resistant manganese steel or nodular cast iron containing chromium carbide that has extremely high abrasion resistance. The crushing hammers are pivoted and thus enabling the mill to allow passage of oversize materials or tramp metal. However, the pivoted hammers exert less force than if they are rigidly attached and are more suitable for soft

FIGURE 3.10   Eriez hammer mill.

Courtesy Eriez (2022c)

material. The mill exit is perforated so that oversize materials are retained and swept up again by the rotor for further crushing. In closed circuit crushing, the discharge from each crushing stage is screened and any oversize is returned to the previous stage for further crushing. Figures 3.10 shows an Eriez hammer mill.[12]

3.5   High Pressure Grinding Rolls (HPGR)

HPGR uses no grinding media but only two counter-rotating tires, one fixed and the other floating to effectively crush ore within a short retention time. As the dry ore feed moves between the two tires in the crushing zone, very high pressure is applied to the system by hydraulic cylinders in a controlled manner leading to inter-particle comminution in the pre-compression and compression zones. The crusher operating gap restricts ore feed size to less than 90 mm. The roll diameter typically ranges from 0.5 to 2.8 m and the width from 0.2 to 1.8 m. Figure 3.11 shows a high pressure grinding rolls.[13]

For the HRC 800 model that weighs about 1 ton, the tire dimensions are 730 × 500 mm, the maximum motor power is 2 × 177 hp, while the height, length and width dimensions are 2400, 3700 and 2700 mm, respectively.

FIGURE 3.11   High pressure grinding rolls.

Source: Courtesy of 911 Metallurgist, 911(f)

3.6   Horizontal Shaft Impact (HSI) Crusher

Horizontal shaft impact (HSI) crusher can perform primary and secondary crushing functions. It is equipped with two jaws, one fixed and the other reciprocating. It operates by applying mechanical pressure on a large rock or ore. The crusher is known for its high crushing reduction ratio, particularly when used for soft and moderately hard materials such as coal, blasted rocks, aggregate, river gravel, mine ores as well as wastes from construction and demolitions.

3.7   Vertical Shaft Impact (VSI) Crusher

The vertical shaft impact (VSI) crusher has been developed to obtain the twister VSI crushers that use two variations of VSI crushing. The first is the VSI crushing in which rock is fed onto a table-type rotor which accelerates and then discharges the rock at high speed against steel anvils in the crushing chamber. The impact of the rock against the steel anvils induces fracture and breaks the rock. The second variation is called vertical shaft autogenous (VSA) crushing, where rock is fed into a multi-port rotor which accelerates and then discharges the rock at high speed against a rock-filled crushing chamber. The impact of rock on rock induces fracture, breaks and shapes the rock. This unique crushing action provides low wear costs, superior products and high reduction ratios.[3]

References

1. Wills, B.A., and Napier-Munn, T.J. (2006): *Mineral Processing Technology*, 7th Edition. Amsterdam: Elsevier Science and Technology Books.
2. Metso (2022): *Basics in Mineral Processing* (https://vdocument.in/basics-in-minerals-processingmetso.html, Accessed 23rd June, 2022).

3. Eriez (2022a): *MACSALAB Jaw Crusher* (www.eriez.com/Images/Product-Images/Size-Reduction-Equipment/Crushing_JawCrushers.jpg?Medium, Accessed 25th April, 2022).

4. Wikimedia (2022): *Jaw Crusher* (https://upload.wikimedia.org/wikipedia/commons/thumb/2/22/Scheme_Jaw_Crusher.gif/320px-Scheme_Jaw_Crusher.gif, Accessed 23rd May, 2022).

5. 911 Metallurgist (2022a): *Jaw Crusher Parts* (https://z4y6y3m2.rocketcdn.me/blog/wp-content/uploads/2021/05/jaw-crusher-parts.png, Accessed 4th May, 2022).

6. 911 Metallurgist (2022b): *Dimensions in a Jaw Crusher* (https://z4y6y3m2.rocketcdn.me/blog/wp-content/uploads/2016/02/crusher_throw-1.png, Accessed 4th May, 2022).

7. 911 Metallurgist (2022c): *Gyratory Crusher* (www.911metallurgist.com/blog/gyratory-crusher, Accessed 16th May, 2022).

8. Eriez (2022b): *MACSALAB Rolls Crushers* (www.eriez.com/Images/Product-Images/Size-Reduction-Equipment/Crushing_RollsCrusher.jpg?Medium, Accessed 25th April, 2022).

9. British Geological Survey (2022): *Cone Crusher* (www.engineeringintro.com/wp-content/uploads/2012/08/cone-crusher.jpg, Accessed 11th May, 2022).

10. 911 Metallurgist (2022d): *Cone Crusher* (www.911metallurgist.com/equipment/laboratory-cone-crusher/, Accessed 11th May, 2022).

11. 911 Metallurgist (2022e): *Impact Crusher* (www.911metallurgist.com/blog/impact-crusher-working-principle, Accessed 30th May, 2020).

12. Eriez (2022c): *Hammer Mill* (https://th.bing.com/th/id/OIP.u-7PeZcqmHZs-EXBefsvNAHaIU?pid=ImgDet&w=60&h=60&c=7, Accessed 30th May, 2022).

13. 911 Metallurgist (2022f): *High Pressure Grinding Rolls* (www.911metallurgist.com/equipment/high-pressure-grinding-rolls/, Accessed 23rd June, 2022).

4 Ore Grinding

4.1 Introduction

Grinding, the last stage in ore comminution can be carried out dry or wet on particles suspended in water and it involves the use of impact and abrasion. The grinding mills comprise cylindrical vessels that carry the ore charge and the grinding mediums which are loose crushing bodies allowed to move freely in the container to break the ore particles. The crushing bodies are typically steel rods, steel balls or rocks of the charge material. There are two types of grinding mills, namely, the tumbling and stirred mills. In tumbling mills such as the rod and ball mills, the motion imparted to the cylindrical vessel is transferred to the charge and the grinding media to effect the comminution. In the stirred mills, on the other hand, a stirrer shaft equipped with pins or discs imparts motion on the charge and the fine grinding media while the mill shell in horizontal or vertical orientation remains stationary.

Tumbling mills are normally used for coarse grinding to grind lumps of size ranging from 5–250 mm to 40–300 μm, respectively. Stirred mills are used to produce fine particles in the fine (15–40 μm) and ultra-fine (<15 μm) ranges, respectively. The principal purpose of the grinding operation is to obtain a correct degree of liberation for the mineral value in the ore. Since each ore has an economically optimum particle size, the grinding mills are variously operated to exercise close control that will ensure the desired product size for each ore. The economic optimum particle size depends on how the mineral value of interest is dispersed in the ore and the separation method the ore will be subjected to.[1]

4.2 Comparison of Crushing and Grinding

There are several differences between crushing and grinding. Crushing is usually carried out dry in several stages with low reduction ratios of between 3 and 6 in each stage. On the other hand, grinding is accomplished wet with much higher reduction ratios to deliver a slurry to the separation processes. Crushing occurs when an ore lump is compressed against rigid surfaces or undergoes impact against rigid surfaces, while grinding takes place when the free movement of the loose, unconnected grinding media causes the exertion of impact and abrasion on the ore lump. Grinding thus typically occurs by tumbling, stirring and vibration.

Another major difference between a crusher and a grinding mill is the smaller retention time in crushing due to its design and this limits the reduction in comparison to grinding. Typical reduction ratios for grinding mills are presented in Table 4.1.[2]

DOI: 10.1201/9781003323433-4

TABLE 4

Reduction Ratios for Selected Grinding Mills

Grinding mill	Ore feed size (mm)	Power rating (kW)	Product size (μm)	Reduction ratio	Mode of operation
Autogenous mill (AG)	400	15–13,000	75	5333.333	Dry/wet
Semi- Autogenous mill (SAG)	400	15–20,000	75	5333.333	Dry/wet
HPGR	90	320–11,400	500	180	Dry
Rod mill	50	3–1500	600	83.333.33	Dry/wet
Ball mill	15	1.5–10,500	20	750	Dry/wet
Vertimill	6	10–3355	5	1200	Dry
Vibrating mill	6	10–75	45	133.333	Dry/wet
Stirred mill	1	18.5–1100	2	500	wet

4.3   The Concept of Correct Grinding

Correct grinding is regarded as a key issue for successful mineral processing for the following reasons:

a. Under-grinding which means grinding to coarse product sizes greater than the liberation size of the mineral value particles and this will negatively affect the efficiency of the mineral processing operation as the degree of mineral value particles liberation may be too low for their separation to occur economically. In addition, it will lead to poor recovery and a low enrichment ratio.

b. Over-grinding implies needlessly grinding ore to a product size much finer than the liberation size of the major mineral value particles in the ore and hence leading to energy wastage in grinding. The energy wastage should be avoided as a report showed that 50% of the energy consumption recorded in the US mills was for comminution. For some Canadian copper ore concentrators, it was also found that the average energy consumed was 2.2, 11.6 and 2.6 kwh/t for crushing, grinding and flotation, respectively and this translates to over 70.7% for grinding processing.

c. The fineness of grinding should not be carried beyond the point where the net smelter return (NSR) due to increased mineral value recovery becomes lower than the additional cost of operation.

d. In hydrometallurgy processes such as gold recovery by cyanidation where leaching recovery of metal value follows the grinding process, further grinding is done to increase the surface area of the ore to be leached even after the mineral value particles have been essentially liberated from the gangue minerals. It has been established that the leaching rate increases with the decrease in feed ore particle size leading to increased recovery of metal values such as gold and this benefits of over-grinding may more than compensate for the increased grinding cost due to increased energy consumption and thus making over grinding justifiable. This may not be applicable to the leaching of base metal ores for which metal recovery may not pay for the increased grinding cost and feeds that are as coarse as possible are the target in such cases.[1]

4.4 Mechanisms of Grinding in Tumbling Mills

Unlike crushing which occurs between relatively rigid surfaces, grinding is a more random process and it is governed by the law of probability. The degree of grinding of an ore particle depends on the probability of the particle getting into a zone within the grinding medium and the probability of some actions taking place after the particle entry into the zone.

Grinding in the tumbling mills can occur by the following mechanisms:

a. Impact or compression due to the forces applied almost normally to the particle's surface
b. Chipping due to oblique forces applied to the particle
c. Abrasion due to forces acting parallel to the surface

The ore particles become distorted and their shapes changed beyond certain limits determined by their elastic limits due to the combined actions of the aforementioned mechanisms. Grinding is mostly carried out wet with the mill charge comprising the ore, the grinding media and water being intimately mixed together and the media breaking the ore particles by any of the mechanisms mentioned depending on the mill shell rotation speed and the shell liner structure. Only a small fraction of the kinetic energy of the tumbling load is expended in particle breakage as most of the energy is lost as noise, heat and others.

4.5 The Motion of the Charge in a Tumbling Mill

The use of loose crushing media such as steel balls, steel rods and rocks is the feature that makes a tumbling mill distinct. These crushing media have the following characteristics:

a. They are large
b. They are relatively hard and high-density materials with respect to the ore charge
c. The mill's volume occupied by these media is typically slightly less than half

The mill's driving force is transferred through the shell lining to the charge. If the speed imparted is low or the shell liner is smooth and thus offers low friction to the ore particles, the grinding medium tends to roll down or undergo cascading to the mill's toe and the abrasion mechanism predominates. This promotes finer grind and higher slime production as well as the wearing of shell lining. At higher speeds, the medium undergoes a cataracting motion that can be described by a series of parabolas before it stops at the mill's toe. This leads to grinding by impact and produces coarser products with lower liner wear. When the medium attains its critical speed, it is projected to have a trajectory profile such that it is expected to land outside the shell. However, centrifuging occurs and the medium will instead be carried around fixed against the mill shell in a position that remains essentially unchanged.

The critical speed of the mill is given by:

$$N_c = \frac{42.3}{\sqrt{D-d}} \, rev \, / \, min \tag{4.1}$$

where
  N_c = mill's critical speed
  D = mill's diameter
  D = the rod or ball medium diameter

Mills are driven in practice at 50–90% of the critical speed based on economic considerations.[1]

4.6   Type of Tumbling Mills

Tumbling mills are of three types, namely ball, rod and autogenous mills. Each type of tumbling mill is made up of a cylindrical shell lying horizontally with an ore charge and an inner wall lined with a renewable-wearing material. The cylindrical drum is carried in hollow trunnions attached to the end walls to be in a position to rotate about its axis. The length and diameter of the mill determine its volume and hence its working capacity, while the mill's diameter determines the force that can be exerted on the charge particles by the medium.

4.6.1   Rod Mills

The grinding media are steel rods. They have reduction ratios of 13–20:1 and thus typically receive feeds with a top size of up to 50 mm and produce ground products as fine as 300 μm. For clayey or damp ores, rod mills are preferred for use over fine crushers that can get choked. The unique feature of rod mill construction is that the length is about 1.5–2.5 times the diameter of the drum. Bond's equation may be used to estimate the power requirements (W) of a rod mill of a particular capacity:

$$W = \frac{10W_i}{\sqrt{P}} - \frac{10W_i}{\sqrt{F}} \tag{4.2}$$

where W_i is the work index for the ore, while F and P are the sizes in μm which 80% of the feed and the product, passes; respectively.

Figure 4.1 shows a typical rod/ball mill setup.[3]
Figure 4.2 shows a typical rod mill shaking tabling system.[4]

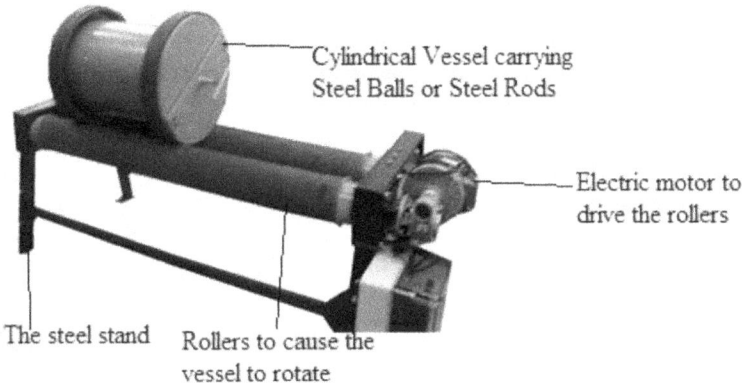

FIGURE 4.1 Eriez MACSLAB rod/ball mill.

Source: **Courtesy of Eriez (2022a), adapted**

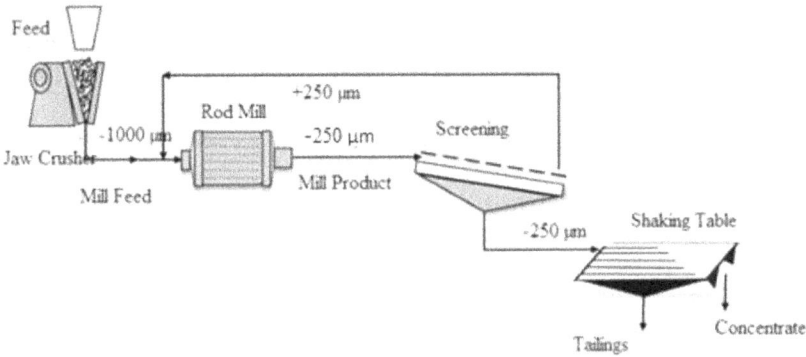

FIGURE 4.2 A rod mill shaking tabling system.

Source: **Adapted from Hamid et al. (2019)**

4.6.2 Ball Mills

Drums that use steel balls as media are used in the final stages of grinding. Balls are considered better suited for fine grinding than rods because on unit weight basis, they have greater surface areas than rods. The term ball mill is restricted to drums with length to diameter ratio of 1.5–1 or less. It has been noted that any degree of fineness can be obtained in ball milling given the required time because ball mills' grinding is by point contact of the grinding balls and particles of the ore. The efficiency of ball mill grinding depends on the following factors:

a. The feed pulp density is required to be as high as possible while taking its ease of flow through the mill into consideration. The thick pulp also ensures ore coating on the balls to avoid metal-to-metal contact that a dilute pulp will cause

b. Depending on the ore type, ball mills are designed to operate with a number of balls that constitute between 65% and 80% of its weight

c. Since pulp viscosity increases with ore fineness, a lower pulp density charge is required in fine grinding circuits

Ball mills for which length to diameter ratio is between 3 and 5 are called tube mills. The grinding medium in tube mills may be rods, balls or pebbles. Tube mills are used to grind gypsum, cement clinker and phosphate. Metso tube mills are designed to grind ore and other materials 35 mesh or finer in both open and closed circuits. Pebble mills typically have long operating lives and require minimum maintenance. They are used for secondary dry or wet grinding. They also typically reduce ores 35 mesh or finer in both open and closed grinding circuits. Figure 4.3 shows a typical ball mill.

The ball mill has become a standard grinding machine in the minerals industry because its operating cost is low and it is simple to construct. The feed into the ball mill typically ranges from 10 to 100 mesh or finer. To produce a ground product whose particles are of uniform sizes, a closed circuit grinding with screening or a classifier is required, for coarse and fine-sized products, respectively. The steel balls are replaced with steel rods for rod milling to obtain products that are almost uniform as well as have the minimum level of fines. A ball or rod mill's maximum power requirement is a function of the grinding medium weight, the design of the liner and the milling speed. Figure 4.3 shows a ball mill assembly.[2, 5]

In a typical industrial dry ball milling practice, the procedure is as follows:

a. The ball mill is filled to half with steel balls, say, weighing 20 kg

b. A 2.5 kg mass of the ore sample is added to the mill

c. The mill is then firmly closed and placed on a roller to be rotated for grinding

5'x10' STEEL-HEAD BALL-ROD MILL WITH DRIVE THROUGH SPEED REDUCER

FIGURE 4.3 A ball mill.

Source: Courtesy of 911 Metallurgist (2022a)

4.7   Stirred Mills

Stirred mills are operated wet and because of their use of smaller grinding media than ball mills, they are ideal for grinding finer products. They are known to be operationally efficient. There are two types of stirred mills, namely, gravity-induced and fluidized technology-based stirred mills. The gravity induced stirred mill initiates a ball charge motion through the rotation of a screw, while the fluidized bed stirred mill carries out size reduction by using a rotational motion to cause a mixture of the grinding media and the slurry to fluidize. Metso South Africa offers Vertimill, a vertical stirred grinding mill that uses gravity-induced technology and stirred media detritor (SMD) vertical fluidized stirred mill.

Stirred mills consume less energy than traditional mills because it is vertical in arrangement, grind by attrition and use finer grinding media. For instance, Metso Vertmill has been reported to achieve energy savings that range from 30% to 50% when compared to the traditional mills, while the energy savings by SMD is more than 50% of ball mills consumption. SMD's ore feeds are typically around 80% passing 100 μm and finer to produce discharge products that can be smaller than 2 μm. The grinding media consist of ceramic or sand of 1–8 mm in diameter.

Stirred mill is also used to treat coarse ore feeds. SMDs can operate on a continuous basis at full load power draw without the product being contaminated with steel chippings. They can be operated in open and closed grinding circuits.

The vibrating mills are operated wet or dry and the grinding occurs by impact, shearing and attrition. It has a short retention time with fewer over-grinding problems. The feed size is <5 mm ore particles. It is typically of low capacity and is used for special applications.

4.8   Autogenous (AG) and Semi-autogenous (SAG) Mills

In these tumbling mills, the ore to be ground itself is used as the grinding media. The ore must contain sufficient competent rock pieces to act as grinding media. In semi-autogenous mills, a combination of steel balls and natural ore grinding media are used.

The AG mill can be operated wet or dry and it is a primary, coarse grinding that can take in up to 400 mm feed size. The grinding media is the grinding feed. AG is of high capacity with a short retention time and it is sensitive to feed composition. The SAG mill can also be operated wet or dry and it has a higher capacity than the AG mill. It can receive ore feed as large as 400 mm in size and is a primary coarse grinding mill like AG. The grinding media is grinding feed plus 4–18% ball charge of diameter 100–125 mm. SAG is of high capacity with a short retention time but it is less sensitive to feed composition, unlike AG.

It has been found that for SAG grinding efficiency, the steel balls should constitute between 4% and 15% of the mill volume. AG and SAG mills are defined by aspect ratio and the nature of the grind product discharge. The aspect ratio is the ratio of the diameter of the mill shell to its length. Based on aspect ratios, there are high aspect ratio, square mills and low aspect ratio mills with aspect ratios 1.5–3 times of length, 1:1 and 0.33–0.66, respectively. AG mills can accomplish the same size reduction work as two or three stages of crushing and screening. The mill reduces the material directly to the desired final product size or prepares it for the subsequent grinding stages. SAG mills have functions similar to AG mills and are ideal for grinding run-of-mine rock or primary crusher discharge.[6, 7]

4.9   Pebble Mill

The pebble mill can also be operated wet or dry and it has a grate discharge. The grinding media comprises of a fraction screened out from feed, flint pebbles, porcelain balls and alumina balls. It is usually larger than ball mills at the same power draw. The advantage is that it grinds without metallic contamination.

The cost of grinding consists of the costs related to energy, liners and the grinding media. The liners used are typically rubber, steel-capped rubber and steel in that order as the severity of the operating condition increases. For AG mills, the total cost of grinding comprises of 37% for liners and 63% for energy, while for SAG mills the cost for liners, grinding media and energy constitute 21, 21 and 58%, respectively.[2, 8, 9]

4.10   Development of Milling Curves

For milling, dry or wet, the time required to grind a sample 80% passing 75 μm for froth flotation can be determined by constructing a milling curve using the following steps:

a. The bulk sample is crushed in a jaw crusher.
b. The crushed sample is passed through a riffler to split.
c. The ore is subjected to grinding for say 5 min based on standard specification.
d. The ground ore is then wet screened through a 75 μm sieve to determine the weight % passing.
e. The procedure in steps 3 and 4 is repeated for other time durations of 10, 15, 20, 25 and 30 min.
f. A plot of % passing 75 μm against time is then constructed.
g. From the plot, the time required to grind for product 80% passing 75 μm can be obtained.

4.11   Open and Closed Circuit Grinding

An open circuit consists of one or more grinding mills, either parallel or in series that discharge a final ground product without classification and no return of coarse discharge back to the mill. The conditions that favor open circuit grinding include:

a. They have small reduction ratios.
b. They reduce particles to coarse, natural grain size.
c. They recycle cleaner flotation middlings for regrinding.
d. They provide non-critical size distribution of the final ground product.

The advantages of open circuit operation over closed circuit operation are:

a. The operation is simple to operate.
b. It is possible to obtain high pulp density discharge.
c. The equipment required is minimal.

Closed circuit grinding consists of one or more grinding mills, either in parallel or in series that discharges milled product to classifiers which in turn return the coarse product from the size separation back to the mill for further grinding. It is the most common grinding circuit in mineral processing because most material and product requirement needs it.

The advantages of closed circuit operation over open circuit operation are:

a. The arrangement usually leads to lower power being consumed per ton of milled product and higher mill capacity.
b. Closed circuit operation leads to avoidance of coarse material in the product by ensuring its return for regrinding.
c. It eliminates over-grinding by removing the fine undersize early enough.

The typically closed circuit arrangement includes Rod mill/Classifier, Ball Mill/Classifier, and Rod mill/Classifier/Ball Mill/Classifier. Figures 4.4 and 4.5 show HPGR and ball mill product and an open circuit ball milling, respectively.[1,4,10]

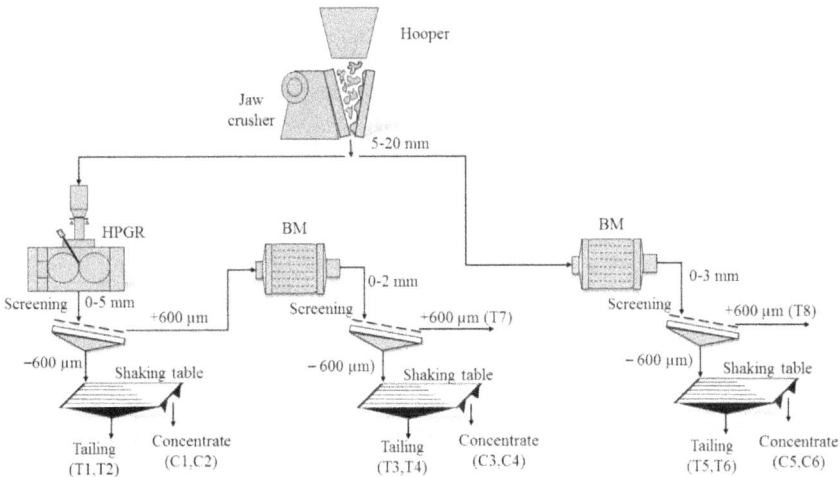

FIGURE 4.4   HPGR and ball mill products.

Source: Adapted from Hamid et al. (2022)

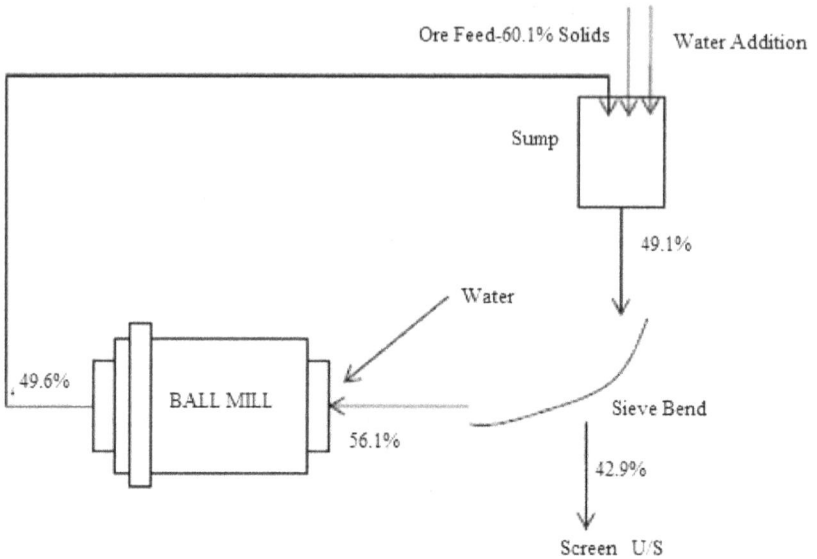

FIGURE 4.5   An open circuit ball milling.

Source: Courtesy of 911 Metallurgist (2022b), adapted

References

1. Wills, B.A., and Napier-Munn, T.J. (2006): *Mineral Processing Technology*, 7th Edition. Amsterdam: Elsevier Science and Technology Books.
2. Metso (2022): *Basics in Mineral Processing* (https://vdocument.in/basics-in-minerals-processingmetso.html, Accessed 23rd June, 2022).
3. Eriez MACSLAB Rod/Ball Mill (www.eriez.com/Images/Product-Images/Size-Reduction-Equipment/RodandBallMill.jpg?Medium, Accessed 25th April, 2022).
4. Hamid, S.A., Alfonso, P., Anticoi, H., Guasch, E., Oliva, J., Dosaba, M., Garcia-Valles, M., and Chagunova, M. (2018): Quantitative Mineralogical Comparison Between HPGR and Ball Mill Products of a Sn-Ta Ore. *Minerals*, 8(4), 151. (https://doi.org/10.3390/min8040151)
5. 911 Metallurgist (2022a): *Ball Mill* (www.911metallurgist.com/blog/ball-mill, Accessed 11th May, 2022).
6. Metso (2020b): (www.metso.com/products/grinding-mills/ag-mills/, Accessed 6th June, 2020).
7. Metso (2020c): (www.metso.com/products/grinding-mills/sag-mills/, Accessed 6th June, 2020).
8. Metso (2020a): (www.metso.com/products/grinding-mills/pebble-mills/, Accessed 6th June, 2020).
9. Metso (2020d): (www.metso.com/products/grinding-mills/stirred-mills/, Accessed 6th June, 2020).
10. 911 Metallurgist (2022b): *An Open-Circuit Ball Milling* (https://www.911metallurgist.com/wp-content/uploads/2016/04/Ball-Mill.png, Accessed 5th April, 2022).

5 Theory of Comminution

5.1 Introduction

Comminution is a sequence of crushing and grinding operations applied to progressively reduce the size of a boulder or a lump. Crushing is applied to reduce the size of the run-of-mine lumps or boulders to the size suitable for grinding. Blasting using explosives can be taken as the first stage in comminution and involves the use of explosives such as mercury, fulminate, a salt of fulminic acid and lead azide $(Pb(N_3)_2)$ to excavate ore from their deposit beds.

The purposes of comminution are:

a. In a quarry, there is a need to obtain aggregate that consists of standard-size fractions or specifications.
b. To make transport of the excavated materials to the plant by scrappers, conveyors and chutes easier.
c. Ore is an aggregate of mineral value and gangue minerals and the mineral value particles are intimately associated with the unwanted gangue minerals. Comminution allows the "un-locking" or "liberation" of the desired valuable mineral(s) from the gangue minerals so that the clean particles of the mineral value are separated from the gangue minerals as much as possible.[1]

5.2 The Theory and Principles of Comminution

Most minerals have their atoms arranged naturally in crystalline form, that is, the atoms are ordered or arranged regularly in a three—dimensional array. The strength of an ore lump depends on the nature of the chemical and physical bonds occurring between the atoms. The bonds between the atoms must be extended until it breaks. A crusher can break a bond strength above 150 MPa. It should be noted that concrete has a bond strength between 20 and 30 MPa. For an ore to be crushed, a stress has to be applied to extend and break the interatomic bond and the stress can be a compressive or a tensile stress. Different minerals have different hardness values on the Mohs hardness scale. Therefore, if stress is applied to an ore, the different minerals in the aggregate will break into fragments with different shapes and sizes. During crushing and grinding, the applied forces may be tensile, compressive, shearing, attrition or impact.

The Mohs hardness scale classifies minerals based on their hardness in ascending order of talc, gypsum, calcite, fluorite, apatite, feldspar, quartz, topaz, corundum and diamond. The hardness levels are such that talc and gypsum can be scratched by a fingernail, calcite by iron, corundum by diamond and diamond being the hardest cannot be scratched by any of the nine other minerals.

DOI: 10.1201/9781003323433-5

In mining and quarrying, rocks and ore minerals are excavated from open and underground mines. The typical operations are drilling and blasting, primary crushing and materials handling. In some natural environments such as glacial, alluvial and marine, primary crushing takes place naturally such as by ice covering, flow of water and natural attrition by erosion leading to sand and gravel of different size distributions. Size reduction is carried out to liberate mineral value from the rock hosting it and thus it is carried out until the liberation size that ranges from 10 to 100 µm is attained. Wearing components endangers the processing machines while noise and dust pose danger to the operators.[1]

Since solid minerals are crystalline, they tend to break into innumerable fragments and shapes upon being subjected to breaking stresses. Consequently, the challenge in ore comminution is to ensure correct grinding that minimizes undergrinding and over-grinding. It has been found that the separation processes become more efficient if the grinding curve for the feed becomes steeper in the latter stages implying shorter or narrower fractions.

The comminution process is influenced by the ore grindability and wear profile, called the work index and abrasion index, respectively. For example, the work and abrasion indices for hematite are 13 ± 8 and 0.5 ± 0.3; respectively. The work index has effects on energy requirement and size reduction, while the abrasion index influences the wear rate of the grinding component.

High reduction ratios lead to in-efficient grinding and hence reduction ratios recommended are 3–4, 3–5 and 3–5 for jaw, gyratory and cone crushers, respectively. For the ball, rod and autogenous (AG)/semi-autogenous (SAG) milling, the reduction ratios recommended are 100, 1000 and 5000, respectively.

It should be noted that the crushing reduction ratios depend on the type of material to be crushed, either rock, gravel or ore such that ore responds with maximum reduction while rock and gravel exhibit limited reduction. Primary, secondary and tertiary crushing typically precede rod and ball milling. The High Pressure Grinding Roll (HPGR) is typically used as a tertiary crusher or quaternary crusher followed by ball milling or vertimilling or as a pebble crusher.[2]

5.3 Comminution Theory

Comminution theory involves the determination of the relationship which exists between the energy input for particle breakage and the particle size produced from a given feed size. The theory can therefore be used to predict the energy requirement for a comminution process. The problems that affect the derivation of a completely satisfactory comminution theory are:

a. For the energy input in crushing and grinding, it has been established that only a small percentage is used for particle breakage; the major part being lost as heat and noise in the machine. Therefore, the small fraction of energy actually expended for particle breakage will be more difficult to determine. For instance, for ball mills, it has been found that less than 1% of the energy supplied for operation is used to actually reduce particle size.

b. It is expected that a relationship should exist between the energy required to break the material to be crushed and the area of the new surface to be produced. However, it is not possible to separately determine the energy expended for creating the new surface and thus the relationship cannot be expressly shown.

c. The comminution theories assume that all ore materials are brittle in nature. However, some ores are plastic in nature and for such ores, energy is expended for shape change which has nothing to do with the actual particle breakage.[1]

5.4   Crack Propagation in Minerals

It has been shown by Griffith (1921)[3] that materials fail by crack propagation when this is energetically feasible, that is, when the energy of the new surface to be created is lower than the energy released when the strain energy created by the comminution is relaxed. Surface energy or interfacial energy or surface free energy quantifies the disruption of intermolecular bonds that takes place when a new surface is created. It can be called the work per unit area done by the force that produces the new surface.

According to thermodynamics, a system tends to proceed in a direction that makes it to be at a lower free energy in the new state so that the change in free energy of the material system is given by $\Delta G < 0$. Brittle materials relieve their strain energy by crack propagation, while in "tough" materials strain energy is relieved by plastic deformation which is shown by a distorting of shape. Comminution theory is an attempt to develop a mathematical relationship between the energy or work input during the crushing/grinding of an ore and the particle size produced expressed in terms of the change in volume, surface area and diameter after the comminution process. The theories of comminution are those of Von Rittinger, Kick and Bond, which were postulated in 1867, 1885) and 1952; respectively.[1]

The Von Rittinger theory states that the energy consumed in size reduction is directly proportional to the area of the new surface created. It should be noted that the total surface area of a given weight of particles of uniform diameter is inversely proportional to the uniform diameter. This implies that the new total surface area increases with the reduction in the diameter of the particles. Rittinger's theory is mathematically stated as:

$$E = K\left(\frac{1}{D_2} - \frac{1}{D_1}\right) \tag{5.1}$$

where
  K = constant
  E = Energy input for the comminution
  D_1 = Initial diameter of ore particles assumed uniform
  D_2 = The final diameter of ore particles assumed uniform

Kick theory states that the energy or work input for size reduction is proportional to the reduction in the volume of the particles concerned. This implies that the reduction in volume increases with the work input exerted. If f and p are the diameters of the feed and product particles, respectively; the reduction ratio R is f/p and Kick's theory can be written mathematically as:

$$E \propto \frac{logR}{log2} \tag{5.2}$$

Bond developed an equation based on the theory that the work input in size reduction is proportional to the new crack tip length produced in particle breakage which is taken to be work represented by the product minus that represented by the feed. The crack tip length is the length of the crack that grows after a fracture occurs rapidly at the same stress. The crack theory postulates that a crack will begin to propagate if the elastic energy released by its growth is greater than the energy required to create the new fractured surfaces.

Bond's equation is given by:

$$W = \frac{10W_i}{\sqrt{P}} - \frac{10W_i}{\sqrt{F}} \tag{5.3}$$

where
  W = work input for crushing/grinding in kilowatts hour per short ton
  P = is the size in μm which 80% of the product particles passes
  F = is the size in μm which 80% of the feed ore passes
  W_i = work index

Therefore, for any rock to be ground, the energy required can be obtained once the Work Index, for the rock is provided. The work index is the comminution parameter which indicates how resistant an ore material is to being crushed and ground. It is numerically the kilowatts hour per short ton to reduce an ore lump from an infinitely large size to 80% passing 100 μm sieve size. One short ton = 907.18474 kg. Work index is also defined as the equivalent amount of energy to reduce one ton of the ore from a very large size to 100 μm. The energy input for grinding in the Bond's test is given by the formula:

$$W = \frac{Mill\,Shaft\,Power}{Mill\,Capacity} = \frac{P_M}{Q} \tag{5.4}$$

The work index can be obtained from plant operations, rod and ball mill grinding tests and impact crushing tests. It is used in the mineral processing industry to compare how resistant different materials are to ball milling. It is used to estimate the energy required for grinding and to design for ball mill scale-up. Ore grindability refers to the ease with which a material can be comminuted and it is widely measured using Bond's work index. Wills and Napier-Munn (2006) reported the

Bond work indices for twelve materials with barite and graphite having work indices of 4.73 and 43.56, respectively. Table 5.1 presents the Bond's Mill Specifications and Grinding Conditions.[1,4,5]

TABLE 5.1
Bond's Mill Specifications and Grinding Conditions

Mill Internal Diameter, Dm, cm	305
Mill Internal Length, Lm, cm	305
Number of Mill Rotations, in minute, n, min−1	70
Mill Balls weight, Mb, kg	20.125
Number of balls	285
Geometry Mill Liner	Smooth
Grinding type	Dry
Volume of material, Vore, cm³	700

TABLE 5.2
Ball Mill Diameters Relationship with Number of Balls for Test

Ball diameter (cm)	No of Balls	Mass (kg)
3.7	43	9.094
3.0	67	7.444
2.5	10	0.694
1.9	71	2.078
1.55	94	0.815

For the mill charge steel balls, the ball mill diameters should be as shown in Table 5.2

In the Bond's standard procedure for work index determination, the following parameters are required:

i. Material: Dry mineral; Size-reduced to 100% < 3350 µm and about 80% ≤ 2000 µm
ii. Quantity: 700 cm³ (tapped down to obtain a reproducible bulk density)

In Bond's standard procedure, the work index is obtained by simulating a ball mill dry grinding in a closed circuit according to Bond's specifications as shown in Table 5.2 such that a 250% circulating load is achieved. The simulation of the circuit grinding is achieved by the constant screening out of the undersize after consecutive batch grinding. The Bond's standard procedure requires 7–10 grinding cycles.

The procedure for the test is outlined as follows:

1. Grind the ball and ore charge mixture dry for 100 revolutions.
2. Screen the ground ore at the starting mesh size of 150 (75 µm) and subsequently at chosen sieve sizes of 106, 150, 212 and 425 µm.
3. Replace the undersize mass with an equivalent mass of the original feed to form a new feed mill.

4. The new mix of fresh feed plus the oversize remainder is ground again as in step 2 but by the number of revolutions that produces a circulating load of 250%, that is, the length of time for each grinding cycle is adjusted until the mass of the oversize fraction is consistently 2.5 times greater the mass of the undersize. This also means that 28.6% of the charge (1/3.5 of total charge) will pass the chosen sieve at the end of the cycle.
5. The procedure is repeated until control sieve undersize produced per mill revolution becomes a constant in the last three milling cycles and reverses its direction of increase or decrease.
6. The final undersize and the new feed are screened to determine the 80% passing size. The 80% passing size of the undersize and the 80% passing size of the new feed as well as the average net weight of the undersize produced per revolution for the last three grinds are subsequently used to calculate the Bond's Work Index with the equation:

$$W_i = \frac{44.5}{P_i^{0.23} G_{bp}^{0.822} \left(\frac{10}{\sqrt{F}} - \frac{10}{\sqrt{P}} \right)} \tag{5.5}$$

where

F = 80% passing size of the new feed before grinding, in µm
P = 80% passing size of the final oversize, in µm
Gbp = average mass, in g, of undersize material produced per revolution for the last three cycles or weight of test sieve fresh undersize per mill revolution (g/rev)
Pi = aperture size of the limiting screen, µm
Wi = Bond Work Index, kwh/t

As a result of Bond's test complexity and the possibility of making mistakes during the complex procedure, several scientists have developed methods to simplify and shorten the procedure. In Berry and Bruce's approximate comparative method, a reference ore of known grindability is used. The reference ore (r) of a given weight is ground for a certain time and the power consumption is noted. The test ore (t) of the same weight is also ground for a length of time such that the power consumed equals that of the reference ore. Then:

$$W_t = W_r = \left(\frac{10W_{it}}{\sqrt{P_t}} - \frac{10W_{it}}{\sqrt{F_t}} \right) = \left(\frac{10W_{ir}}{\sqrt{P_r}} - \frac{10W_{ir}}{\sqrt{F_r}} \right) \tag{5.6}$$

Therefore

$$W_{it} = \frac{W_{ir}\left(\frac{10}{\sqrt{P_r}} - \frac{10}{\sqrt{F_r}} \right)}{\left(\frac{10}{\sqrt{P_t}} - \frac{10}{\sqrt{F_t}} \right)}$$

where

W_{ir} = *known work index of the reference ore*

W_{it} = *work index of the test ore to be determined*

P_r = the size in μm which 80% of the reference ore product passes

P_t = the size in μm which 80% of the test ore product passes

F_r = the size in μm which 80% of the reference feed ore passes

F_t = the size in μm which 80% of the test ore feed passes

It has been noted that reasonable values of work indices are only determined from this method if about the same product size distribution is obtained for the reference and test ores ground under the specification.

5.5 Worked Examples

Example 5.1

It is proposed to reduce 21, 000 tons of dolomite from 80% passing 425 μm to 80% passing 200 μm over a period of one week (7 days). If there are three jaw crushers of the same type arranged in parallel to be used, determine:

a. The tonnage output of each jaw crusher per hour

b. The total power required for the primary jaw crushing

Take the Work Index for dolomite as 12.44 kwh/t.

Solution

a. 21,000 tons is to be crushed in 7 days

The quantity to be crushed per day $= \dfrac{21000}{7} = 3000$ tons / day

The quantity to be crushed per hour (I) = 3000/24 = 125 tph

Each crusher is then required to crush $= \dfrac{125}{3} = 41.67$ tons / h

This means tonnage per crusher = 1000.08 tons/day

b. The total power required for the crushing

$$P = 10 W_i I \left| \frac{1}{\sqrt{P_{80}}} - \frac{1}{\sqrt{F_{80}}} \right|$$

But: $W_i = 12.44$ kwh / ton for dolomite,

I = P80 = 200 μm, F80 = 425 μm

$$P = 10 \times 12.44 \times 125 \times \left| \frac{1}{\sqrt{200 \times 10^{-6}}} - \frac{1}{\sqrt{425 \times 10^{-6}}} \right|$$

$$= 10 \times 12.44 \times 125 \times (70.71068 - 48.50713) = 345265.3\,\text{W}$$
$$= 345.27\ \text{kW}$$

The power required for each crusher = 115.09 kW.

Example 5.2

One kg of a new ore was ground in a rod mill with a 22 kg charge of rods that weighed 20 kg for 10 min. It was found that 80% of the ore as received passed the 1.56 mm screen size, while 80% of its ground product passed the 120 μm screen. The same rod mill was used to grind ore with a known work index of 12.86 kwh/t with the same sizes and weights of rods for the same 15 min. The results obtained showed that the ore with 80% passing 1.20 mm as received gave a product with an 80% passing size of 78 μm. Determine, using the Bond Work Index formula, the work index of the first ore.

Solution

Sample 1
F80 = 1.56 mm = 156 × 10^{-2} mm
But 1000 mm = 1m
F80 = 156 × 10$_{-2}$ × 10^{-3} m = 156 × 10^{-5} m = 156 × 10 × 10^{-1} × 10^{-5} m
    = 1560 × 10^{-6} m = 1560 μm
P80 = 120 μm
Sample 2
F80 = 1.20 mm = 1200 μm
P80 = 78 μm
Wi = 12.86 kwh/t
Since the grinding in both cases took place for 15 min, the work done in both cases was the same. Therefore:

$$W_1 = W_2$$

$$10W_{1i}\left|\frac{1}{\sqrt{P_{80,1}}} - \frac{1}{\sqrt{F_{80,1}}}\right| = 10W_{2i}\left|\frac{1}{\sqrt{P_{80,2}}} - \frac{1}{\sqrt{F_{80,2}}}\right|$$

$$W_1 = 10W_{1i}\left|\frac{1000}{\sqrt{120}} - \frac{1000}{\sqrt{1560}}\right|$$

$$W_2 = 10W_{2i}\left|\frac{1000}{\sqrt{78}} - \frac{1000}{\sqrt{1200}}\right|$$

$$W_2 = 10 \times 12.86 \times \left|\frac{1000}{\sqrt{78}} - \frac{1000}{\sqrt{1200}}\right|$$

$$10W_{1i}\left|\frac{1000}{\sqrt{120}}-\frac{1000}{\sqrt{1560}}\right|=10\times12.86\times\left|\frac{1000}{\sqrt{78}}-\frac{1000}{\sqrt{1200}}\right|$$

$$W_{1i}=\frac{10\times12.86\times\left|\dfrac{1000}{\sqrt{78}}-\dfrac{1000}{\sqrt{1200}}\right|}{10\left|\dfrac{1000}{\sqrt{120}}-\dfrac{1000}{\sqrt{1560}}\right|}$$

$$W_{1i}=\frac{12.86\times\left|\dfrac{1}{\sqrt{78}}-\dfrac{1}{\sqrt{1200}}\right|}{\left|\dfrac{1}{\sqrt{120}}-\dfrac{1}{\sqrt{1560}}\right|}=\frac{1.084872}{0.065969}=16.45\,\text{kwh/ton}$$

Example 5.3

Consider the crushing of ore with an F80 of 420 mm and a P80 of 20 mm as rod mill input. If the reduction ratios for primary and secondary crushings are 3 and 4, respectively, determine the number of crushing stages required for the ore.

Solution

The F80 = 420 mm
This means that 80% of the feed material passes a 420 mm sieve
The P80 = 21 mm
This means that 80% of the rod mill input material passes a 21 mm sieve
The overall or total reduction ratio required for the ore-crushing process =
$$\frac{F80}{P80}=\frac{420}{21}=20$$

But, R1 = 3 and R2 = 4
Therefore, the total reduction ratio attainable for a primary crushing followed by a secondary crushing = R1 × R2 = 12
Since this is less than 20, an additional crushing stage is required.
For instance, using one stage of jaw crusher primary crushing and two stages of cone crusher secondary crushing will give a total reduction ratio = R1 × R2 × R2 = 3 × 4 × 4 = 48 >> 20

However, when using several crushing stages, the practice is to make the process more flexible by reducing the reduction ratios. For example, R2 can be reduced to 3 to obtain the total reduction ratio =

$$3\times3\times3=27>21$$

This three-stage crushing will thus give sufficient reduction
There are also mobile crushers that combine crushing and screening capacities such as jaw + grizzly and impact + grizzly mobile crushers.

References

1. Wills, B.A., and Napier-Munn, T.J. (2006): *Mineral Processing Technology*, 7th Edition. Amsterdam: Elsevier Science and Technology Books.
2. Metso (2022): *Basics in Mineral Processing* (https://vdocument.in/basics-in-minerals-processingmetso.html, Accessed 23rd June, 2022).
3. Griffith, A.A. (1921): The Phenomena of Rupture and Flow in Solids. *Philosophical Transactions of the Royal Society*, 221, 163.
4. Todorovic, D., Trumic, M.S., Andric, L., and Trumic, M. (2017): A Quick Method for Bond Work Index Approximate Value Determination. *Phsicochemical Problems in Mineral Processing*, 53(1), 321–332.
5. Man, Y.T. (2002): Why Is the Ball Mill Grindability Test Done the Way It Is Done? *The European Journal of Mineral Processing and Environmental Protection*, 2(1), 34–39, 1303–0868.

6 Particle Size Analysis

6.1 Introduction

Particle size analysis refers to the analysis of the shape, size and size fraction distribution range of ore at every stage of mineral processing such as after crushing, grinding, separation and product collection. After crushing and grinding, the aggregate of ore minerals with different Mohs hardness values and locations within the ore lump will break into different sizes and shapes. In particle size analysis, the range of sizes and shapes produced depending on the severity of the comminution process will be determined.[1]

6.2 Importance of Particle Size Analysis in Mineral Processing

Particle size analysis or screen distribution analysis or sieve analysis is an important procedure in mineral processing for the following reasons:

a. It enables one to know the quality of grinding carried out on an ore using the percentage undersize at a particular size as an indicator.
b. It enables one to know the extent of liberation of mineral value particles from the gangue mineral particles at different size fractions. A microscopic analysis or size by assay of the size fractions after crushing or grinding will reveal the degree of liberation of the mineral value at each size fraction.
c. It provides the optimum ore feed size consist that will give us the maximum efficiency in the mineral processing operation. Depending on the Moh hardness of each mineral type in the ore, the crushed product will contain different particle sizes and shapes after crushing depending on the initial size of the lumps.
d. It enables knowing the size ranges fractions at which any losses can occur in mineral processing operation so that such losses can be reduced.

Since major plant decisions are made on a routine basis using the results of particle size analysis, two important issues about particle size analysis have to be ensured:

a. The method applied should be reliable and accurate.
b. The sample used must be a representative of the bulk original (bulk) sample as accurately as possible.

6.3 Particle Size and Shape

The primary purpose of precision particle size analysis of ore is to obtain quantitative data about the size and size distribution of mineral particles in it. Ore particles can occur in regular shapes such as spheres or cubes or irregular shapes. For regular shapes, the exact size can be uniquely assigned while for irregular shapes it is not

DOI: 10.1201/9781003323433-6

possible to assign them an exact size because "breadth", "length", "width", diameter or "thickness" dimensions cannot be uniquely determined unlike for regular particles. For instance, for spherical and cubical particles, the exact sizes are the diameter and the length of a side, respectively.

The size of an irregular particle can be stated in terms of the parameter called the "equivalent diameter" which refers to the diameter of a sphere that would when subjected to some specified operations exhibit the same behavior as the irregular particle. Size analysis should be accompanied by a remark about the approximate shape of the ore particles. There are different equivalent diameters depending on the method of measurement for carrying out the specified operation, namely:

a. Stokes' diameter that uses sedimentation and elutriation for measurement.
b. The Projected Area diameter which is measured using a microscope.
c. The Sieve Aperture diameter which is measured by test sieving. The width of the aperture of a sieve through which the irregular particle just passes is taken as its equivalent diameter. For irregular particles, the equivalent diameter in this case refers to their second-largest dimension, not the largest.
d. The approximate shape of ore mineral particles after comminution can be
    i. Spherical shape
    ii. Granular—with approximately an equidimensional irregular shape
    iii. Global shape
    iv. Modular—rounded, irregular shape
    v. Acicular—needle shape
    vi. Dendritic—having a branched crystalline shape
    vii. Fibrous shape—regular or irregular thread-like
    viii. Flaky—plate-like
    ix. Irregular—lacking any symmetry
    x. Angular—sharp-edged

6.4   Methods of Particle Size Analysis

Some of the methods of particle analysis are:

1. Test sieving—this is suitable for both dry and wet analysis
2. Laser diffraction—this is also suitable for both dry and wet analysis
3. Optical microscopy—this is suitable for only dry analysis
4. Electron microscopy—this is suitable for only dry analysis
5. Elutriation—this is suitable for only wet analysis
6. Sedimentation (gravity)—this is suitable for only wet analysis
7. Sedimentation (centrifuge)—this is suitable for only wet analysis

6.4.1 Method of Test Sieving

Important facts to note about test sieving are:

1. Test sieving is an old method of particle size analysis and it is done using sieves with size specifications based on different standards. The different standards with different size ranges are:
   a. The American Standard for Testing and Materials (ASTM)
   b. The German Standard known as DIN 4188
   c. The British Standard known as BS 1796
   A given mass of ore material is passed through a set of an agitated stack of sieves arranged in descending order of aperture size and the material retained on each sieve is weighed to determine the percentage distribution. The agitation of the sieves is done to ensure equal exposure of all sample particles to the sieve aperture openings.
2. Designation of sieves
   Woven wire sieves are designated based on the nominal aperture size. They can be defined in two ways
   d. Nominal central separation between opposite sides of a square aperture or the nominal diameter of a round aperture. This is the preferred modern designation
   e. Number of wire per inch or number of squares per square inch which is called mesh number
3. Mechanism of sieving

The following should be noted:

a. The amount of sample used should not be too much as this will increase the depth of the sample in the sieve and adversely affect the favorable orientation of some particles to the sieve opening for passage. The sample's weight should also not be too low so that it contains a sufficiently large number of particles to represent the original bulk adequately.
b. The particles in the sample can be categorized into three types on exposure to an aperture opening, namely, the undersize particles that readily pass through the aperture opening on favourable orientation to it, the oversize particles that cannot pass the aperture and the near size particles that cause sieve blinding and whose passage, if it occurs, is a gradual process. Figure 6.1 shows the Eriez MACSALAB sieve shaker with mounted sieves.[2]

The sieve shaker unit shown earlier has a heavy cast iron unit with a speed regulator that ensures a uniform vibration on a sieve table. The motor actuates an eccentric weight that will produce translation and rotation. Table 6.1 presents the selected American Standard test sieve series.[3]

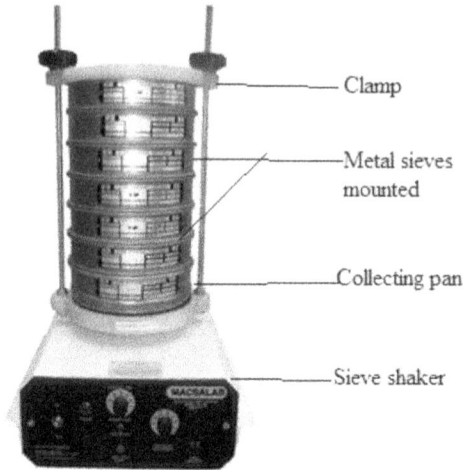

FIGURE 6.1 Eriez MACSALAB sieve shaker with mounted sieves.

Source: Courtesy of Eriez (2022)

TABLE 6.1
Selected American Standard Test Sieve Series

Standard	Alternative	Standard	Alternative
125 mm	5.00	180 μm	No. 80
106 mm	4.24	150 μm	No. 100
100 mm	4.00	125 μm	No. 120
90 mm	$3\frac{1}{2}$	106 μm	No. 140
75 mm	3.00	90 μm	No. 170
63 mm	$2\frac{1}{2}$	75 μm	No. 200
850 μm	No. 20	63 μm	No. 230
350 μm	No. 45	53 μm	No. 270
300 μm	No. 50	45 μm	No. 325
250 μm	No. 60	38 μm	No. 400
212 μm	No. 70	32 μm	No. 450

6.4.2 Test Sieving Procedure

The test sieving involves the following steps:

1. Select your sieves from the range available in the standard you are using. Select say 9 sieves using the square root of two methods such that $\frac{S1}{S2} = \sqrt{2}, \frac{S3}{S4} = \sqrt{2}, \frac{S4}{S5} = \sqrt{2}$ and so on such that S1, S2, S3, S4 and S5 are

the aperture sizes for sieve numbers 1, 2, 3, 4 and 5, respectively. For example, if the ore sample is such that the coarsest sieve required based on its top size is 600 μm, then the next sieve will be obtained from $\dfrac{S1}{S2} = \dfrac{600}{S2} = \sqrt{2}$ so that $= \dfrac{600}{\sqrt{2}} = 424.3$ and thus the succeeding sieve will be of size 424 μm or approximately 425 μm since there is no standard sieve with size 424 μm.

2. Take a known weight of the representative sample for the analysis according to the guidelines stated earlier.
3. Weigh each of the nine sieves and the pan empty (m_i).
4. Stack the sieves as a nest, place and clamp it on the sieve shaker.
5. Switch on the sieve shaker machine to agitate the clamped stack of sieves for a number of minutes, say 20–30 min.
6. Take the sieves apart with the fractions retained on each.
7. Then, weigh each of the sieves with the retained fractions on each (m_f).
8. Finally, determine the mass retained on each sieve by subtracting the weight of the sieves in (3) from that in (7), that is, $m_f - m_i$.

6.4.3   Presentation of Results

The results of sieve tests carried out on a Nigerian gold ore sample ore is presented in Table 6.2.[4]
The following can be observed from Table 6.2:

1. Column 1 shows the range of sieves used in the test sieving, the coarsest sieve being 1180 μm and the finest 53 μm.

TABLE 6.2
Particle Size Distribution of the As-received Gold Ore Sample

S/N	Sieve Size (μm)	Retained Weight (g)	Cumulative Oversized (g)	Percentage Retained (%)	Cumulative Percentage oversize (%)	Cumulative Percentage undersize (%)
1	+1180	11.4	11.4	3.80	3.80	96.20
2	−1180+850	16.3	27.7	5.43	9.23	90.77
3	−850 + 600	14.3	42.0	4.77	14.00	86.00
4	−600 + 425	50.2	92.2	16.73	30.73	69.27
5	−425 + 300	41.5	133.7	13.83	44.57	55.43
6	−300 + 212	64.1	197.8	21.37	65.93	34.07
7	−212 + 150	65.4	263.2	21.80	87.73	12.27
8	−150 + 106	26.3	289.5	8.77	96.50	3.50
9	−106 + 75	6.9	296.4	2.30	98.80	1.20
10	−75 + 53	3.0	299.4	1.00	99.80	0.20
11	−53	0.6	300.0	0.20	100.00	0.00

2. Column 3 shows the weight of material retained on each sieve such that an 11.40 g fraction of the sample mass was retained on the 1180 μm sieve, while a 16.30 g fraction of the iron ore sample passed the 1180 μm sieve but was retained on the succeeding 850 μm sieve.
3. Column 4 shows the weight percentages of the ore retained on each sieve. Columns 6 and 7 give the cumulative weight percentages of oversize and undersize materials, respectively
4. Rows 10 and 11 show that 3.0 g of ore was retained on the last sieve 53 μm, while 0.60 g of the ore passed the sieve and was collected in the pan below it.

The newer techniques of particle size analysis have been found to provide improved accuracy, reliability and reduced time for analysis. The sieve apertures in test sieves can also be mechanically altered and worn with use and damaged sieve mesh will yield misleading results and thus the justification for the new techniques of particle size analysis developed.

6.4.4   Construction of Test Sieve Graphs

Test sieving results are always presented on a graph to evaluate their significance. The methods to plot the results include the following:

1. Plotting of the nominal aperture size in μm against the cumulative undersize or oversize in percentage. The normal arithmetically ruled graph paper can be used but it leads to a congestion of the graph line at the finer sieve range. In view of this, a semi-logarithmic graph where the percentage undersize or oversize is on the linear ordinate and the nominal aperture size is on the logarithmic abscissa is preferred.
2. The cumulative percentage undersize is a mirror image of the cumulative percentage oversize and thus it is not necessary to plot the two.
3. Many of the plots of cumulative percentage undersize or oversize against particle size produce S-shaped curves and this lead to data congestion at the two ends. Other types of graphs to present the test sieve results in order to avoid this drawback include the following:
   a.   The Gates-Gaudin-Schuhmann method graph in which undersize are plotted against nominal particle size on log-log graph paper
   b.   The Rossin-Rammler method graph presents the particle size distribution of the ball mill product which can be represented by the equation:

$$100 - P = 100\exp\left(bd^{n}\right) \tag{6.1}$$

where
   P = cumulative undersize in percentage
   d = the particle size
   b, n are constants

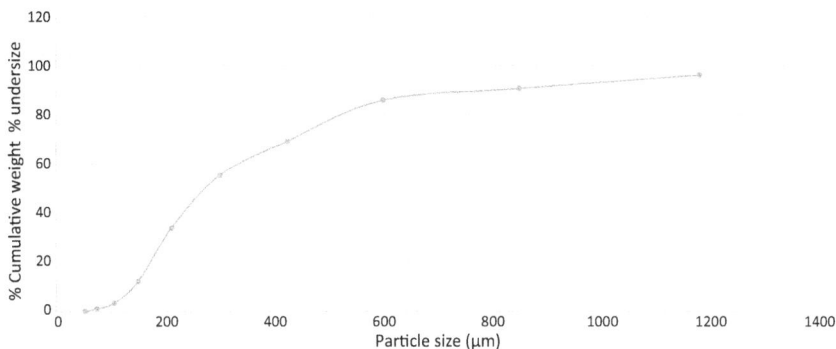

FIGURE 6.2a   Cumulative per cent undersize curve.

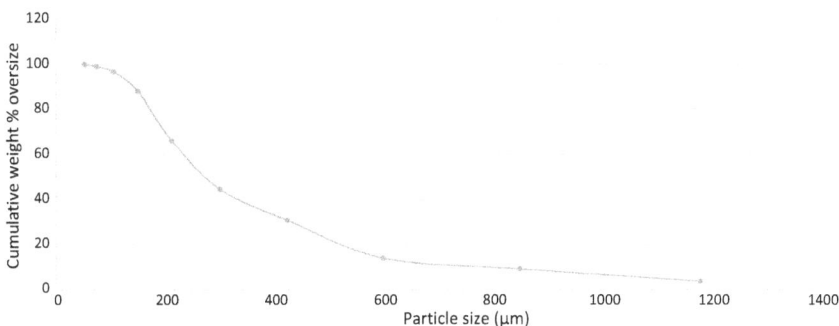

FIGURE 6.2b   Cumulative per cent oversize curve.

Figures 6.2a and 6.2b present the cumulative per cent undersize and oversize curves, respectively.

The mean of a size fraction interval is obtained by taking the average of the sieve apertures of two succeeding sieves. For instance, for the size fraction on row number 2, the mean is 1015 μm.

6.4.5   Deductions from the Curves

The following deductions can be made from the screen distribution curves:

a. The mid-point size of the sample particles, called the median size of the sample, such that 50% of the particles are of sizes larger and 50% are of sizes smaller.

b. Assessment of the grinding circuits' performance. The ground ore produced is often specified by the 80% passing size point on the cumulative per cent undersize, which is the point at which the cumulative per cent undersize is 80%. For instance, the target size can be stated as 80% <150 μm. This means that the ore feed should be ground such that 80% of the product passes the 150 μm sieve aperture. If the ground product is screened through the 150 μm sieve and found to contain 60% particles less than 150 μm, the ground product is coarser than the specifications and the product has to be ground more.

c. From the curves, the cumulative weight percentage undersize can be deduced for any nominal aperture size.

6.5 Determination of Stokes' Equivalent Diameter

Both sedimentation and elutriation methods are based on the resistance of the ore particles to movement in a fluid. In the sedimentation method, the ore material to be screen sized is dispersed in a fluid in which the particles settle in response to the forces such as those due to gravity, centrifugal acceleration or electromagnetism acting on them. In elutriation, the ore particles, usually smaller than 1 μm, are separated or sized in a stream of gas or liquid flowing in a direction opposite to the direction of sedimentation based on their size, shape and density. In sedimentation, the resistance to the particle's motion as it falls under gravity determines the terminal velocity each particle will attain.

For spherical particles with sizes <38 μm, that is, particles within the sub-sieve range, the terminal velocity can be predicted with Stokes' equation:

$$V = \frac{d^2 g \left(D_s - D_f \right)}{18\eta} \qquad (6.2)$$

where
υ = the ore particle's terminal velocity (m/s)
d = the ore particle's diameter (m)
D_s = the ore particle's density (kg/m³)
D_f = density of the carrier fluid in which the particle is suspended (kg/m³)
υ = viscosity of the carrier fluid (N/s/m²)
g = acceleration due to gravity (m/s2)

A non-spherical ore particle will also attain a terminal velocity but the velocity will be affected by its shape and so cannot be correctly predicted by Stokes' equation. However, if the terminal velocity of a non-spherical particle is known and it is substituted in Stokes' equation, the diameter d obtained will be Stokes' equivalent spherical diameter or Stokes's diameter for the ore particle.

6.6 Dynamic Image Analysis

This method of image analysis applies modern cameras and software to samples with ultra-fine particle sizes only larger than several microns in diameter. Typical examples of dynamic image analyzers are the CAMSIZER P4 and CAMSIZER X2.

The image analyzer takes a photograph of each particle and analyse the resulting photo to determine particle size and shape.

The shape and size of irregular granular materials are important as they are the basic physical properties that define them and determine some of their properties such as flatness and strength. They also influence their flow parameters in liquids such as sharpness of separation and rate of sedimentation. Grains are considered flat if the ratio of length to thickness exceeds 3. Flatness leads to disturbances in a particle's flow and negatively affects structural strength. The shape of a particle is characterized using 1D, 2D and 3D linear, circular and spherical shape descriptors, respectively that use the dimensions such as length, diameter, circumference, surface area or volume. For instance, a 1D shape descriptor will use linear shape parameters such as flatness, while a 3D one will depend on the 3D space of the particle such as its spherical shape factors like volume and surface area. By convention, sphericity is defined as the ratio between the surface area of a sphere of the same volume as the particle and the actual surface area of the particle.[5]

6.7   Laser Diffraction

A laser is a device that emits light through a process of optical amplification based on the stimulated emission of electromagnetic radiation. The term "laser" is an acronym for "Light Amplification by Stimulated Emission of Radiation". Depending on the source, there are solid-state, liquid and semiconductor lasers. Lasers are ultraviolet radiations with a wavelength between 180 and 700 nm.

As particle size decreases, sieving and/or dynamic image analysis becomes more difficult or even impossible and laser diffraction becomes the technique of choice. Once the particles are too fine for image analysis—roughly several microns—they are best analyzed by laser diffraction instruments such as the Horiba LA-960 and LA-300. Laser light is passed through a dilute suspension of the particles which circulate through an optical cell. The suspended particles scatter the laser light and are detected by a solid state device which measures the scattered light intensity over a range of angles. The particle size distribution is calculated from the light distribution pattern based on the theory of light scattering such that finer particles produce more scatter than coarse particles. The early instruments were based on the Fraunhofer diffraction theory which is suitable to describe coarse particles in the range of 1–2000 µm. Mie theory was later introduced to improve the capability to ultra-fine particles 0.1 µm and below.

The Fraunhofer diffraction and Mie light scattering theories are used to determine what type of light distribution patterns are produced by various particle sizes and are stored on a computer as a parameter table. The Fraunhofer diffraction is one of the approximations of the Mie light scattering theory and can only be used when the particle size is relatively large, at least ten times the laser wavelength and when the scattering angle is less than 30°. Figures 6.3 and 6.4 show a Laser diffraction analysis equipment and a diffraction analysis image 30°.[6-9]

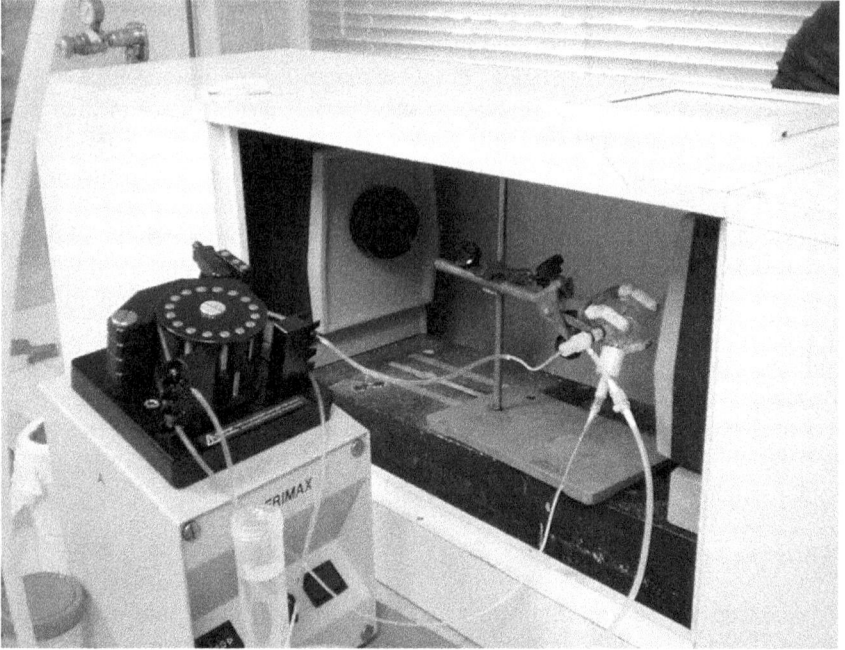

FIGURE 6.3 Laser diffraction analysis equipment.

Source: Courtesy of Wikimedia (2022a)

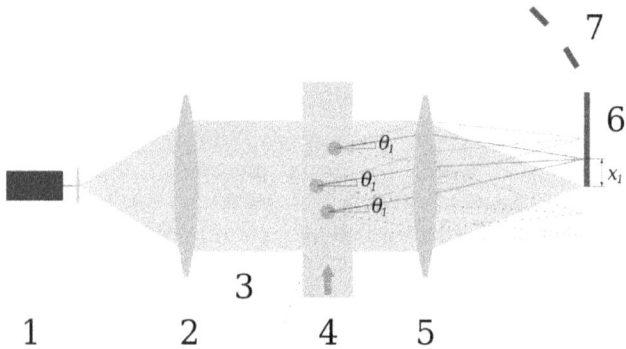

FIGURE 6.4 Sketch of LASER diffraction analysis.

Source: Courtesy of Wikimedia (2022b)

6.8 Determination of Particle Shape and Size

A LEICA Galen III Research Microscope with an integrated camera (Celestron digital microscope imager, model 44421) is used to view an ore sample and is typically carried out as follows. The microscope is set in position and switched on. About 0.5 g of 80% passing 75 µm of the ore is evenly spread on a glass slide and placed on the objective section of the microscope. The microscope is then set to X100 magnification and the adjustment knob is manipulated until a sharp image is obtained. Picture of the sharp image is then taken. The photomicrographs taken are analyzed using Rasband Image–J software to determine particle parameters such as aspect ratio, elongation ratio and roundness.

References

1. Wills, B.A., and Mapier-Munn, T.J. (2006): *Mineral Processing Technology*, 7th Edition. Amsterdam: Elsevier Science and Technology.
2. Eriez (2022): *Macsalab ES — 200 220 Volt Electrical Sieve Shaker* (www.eriez.com/Images/Product-Images/Size-Reduction-Equipment/SieveShakers.jpg?Medium, Accessed 25th April, 2022).
3. Endocotts (www.endecotts.com/faq/sieves-and-calibration/american-standard-test-sieve-series-astm/, Accessed 4th August, 2020).
4. Teniola, O.S. (2023): *Production of High Concentrate Gold Solution from Aqua-Regia Leachate Using Bagasse Nanoparticles*. A PhD Thesis, Department of Materials Science and Engineering, Obafemi Awolowo University, Ile Ife, Nigeria.
5. Horiba (www.horiba.com/en_en/applications/materials/metal-and-mining/particle-analysis-in-mining-and-minerals/, Accessed 13th June, 2020).
6. Schimadzu Excellence in Science, Global Analytical and Measuring Instruments (www.shimadzu.com/an/powder/support/practice/p01/lesson13.html, Accessed 18th June, 2020).
7. Krawczykowski, D., Krawczykowska, A., and Gawenda, T. (2022): Comparison of Geometric Properties of Regular and Irregular Mineral Grains by Dynamic Image Analysis (2D) and Optoelectronic Analysis (3D) Methods. *Minerals*, 12, 540. (https://doi.org/10.3390/min12050540)
8. Wikimedia (2022a): *SizeDistributionAnalysisEquipment* (https://upload.wikimedia.org/wikipedia/commons/thumb/6/68/Drop_size_distribution_analysis_4.jpg/800px-Drop_size_distribution_analysis_4.jpg) By Superchilum—Own work, CC BY-SA 3.0 (https://commons.wikimedia.org/w/index.php?curid=16163321, Accessed 24th May, 2022).
9. Wikimedia (2022b): *Sketch of Diffraction Analysis* (https://upload.wikimedia.org/wikipedia/commons/thumb/1/19/Laser_diffraction_analysis_sketch.svg/1024px-Laser_diffraction_analysis_sketch.svg.png, Accessed 24th May, 2022) By Sunspeanzler—Own work, information source e.g. Laser Diffraction, product information, Company Sympathec GmbH, CC BY-SA 3.0 (https://commons.wikimedia.org/w/index.php?curid=35816080; https://en.wikipedia.org/wiki/Laser_diffraction_analysis).

7 Ore Mineralogy and Chemical Analyses

7.1 Ore Mineralogy

Mineralogy is the study of minerals and their physical and structural properties. For efficient processing of an ore, it is necessary to know the mineralogy of the ore thoroughly. Mineralogy test is done with a microscope and it enables us to have the following information[1]:

(1) The nature of the mineral values and the associated gangue minerals
(2) The texture of the ore which includes the following:
    (a) The size of the mineral grains
    (b) The shape of the mineral grains
    (c) Mineral value dissemination in the gangue
    (d) Association between the mineral value(s) and the gangue minerals
(3) The potential difficulty that may be encountered in separating the mineral value(s) from the gangue minerals
(4) The production of possible or feasible concentrate grade
(5) The analysis of the concentrate and tailings microscopically also enables one to know how efficient the liberation and separation processes are

The methods of microscopy used are as follows.

Visual Inspection

Many minerals can be identified by inspecting them visually and performing some simple tests. There are about 4000 known minerals, the basic category being the rock-forming minerals. Rock-forming minerals are minerals that constitute the earth's crust rocks and they include quartz, mica, calcite and feldspar. The chemical formulae of orthoclase feldspar, mica lepidolite and calcite are $KAlSi_3O_8$, $K(Mg, Fe)_3(AlSi_3O_{10})$ and $CaCO_3$, respectively. Most rock-forming minerals are silicates as seen from their chemical formulae with feldspar constituting over 41% of the continental crust. Rocks often contain several minerals with some present in small concentrations and either mechanically or chemically combined with other minerals.

In the visual identification of rock-forming minerals, very small grains or larger hand specimens that can easily be readily observed and manipulated can be used. In identifying metallic and rock-forming minerals, the use of a field guide that contains features such as image, color, crystal shape and lustre is important. Table 7.1 shows an example of a field guide used in identifying and differentiating between sulphur, zirconia, hornblende and galena.[2]

DOI: 10.1201/9781003323433-7

TABLE 7.1
Characteristics of Some Common Minerals

Mineral	Sulphur	Zirconia	Hornblende (Calcium amphibole)	Galena
Color	Bright yellow as crystals; pale yellow as powder	Colorless, black, blue, green, yellow, red or brown	Black to dark green	Lead or silver grey; may have a bluish stint
Crystal shape	Orthorhombic	Hexagonal	Monoclinic	Octahedral
Lustre	Glassy to earthy	Adamantine	Glassy to dull	Metallic to dull

Conventional optical microscopes are also used to observe thin and thick polished surfaces. In thin-section microscopy, the sample is ground to about 30 μm thickness so that it is transparent to light.

7.2 Transmitted Light Microscope

A light microscope is also called an optical microscope. The microscope has lenses that are used to focus light on samples placed close to magnify them and then take their images. Microscopic magnification depends on the type and the number of lenses it contains. A simple microscope has a low magnification because it has only one lens, while a compound microscope has higher magnification as it has a minimum of two sets of lenses-the objective lenses and an eyepiece. The lenses are aligned in such a way that they can bend light to efficiently magnify images. A thin sample is transparent, that is, it allows the beam of focused light to pass through it to generate an image which is then taken through one or more lenses for magnification and viewing. On the other hand, an electron microscope focuses electrons on specimens using magnets and produces much greater magnifications since electrons have shorter wavelengths.

The polarizing transmitted-light microscope, also called the petrographic microscope, is basically made up of a light source, a sub-stage condenser, a specimen holding stage, an objective and an eyepiece. Furthermore, the microscope has a polarizer, a rotating graduated stage and a second polarizer called an analyzer. The light source is a lamp built at the base of the microscope. It has been postulated that light such as that from a microscope bulb is made up of electromagnetic vibrations that move outwards in every direction from the point source. The light is passed through a polarizer below the microscope stage and becomes polarized and vibrates along a single plane such as East-West and is called a plane polarized light (PPL). One or two field diaphragm is located below the stage to reduce the area of the incident light entering the thin section. The condenser or convergent lens is attached to a swivel bar and it directs light of a conical shape to the thin section to ensure optimum resolution.

The stage is flat and can be rotated and the sample rock thin section is attached to the stage's centre by metal spring clips. Objectives are magnifying lenses with magnification power such as × 30 inscribed. The analyzer receives the light vibrating in the East-West from the polarizer but it cannot transmit it and thus produce a dark field of view colors, while a bright field of view is obtained without the analyzer. The moving of the microscope stage up and down as well as the use of coarse and fine adjusting knobs are used to focus the microscope. The light microscopes in a polarized state can be used in the reflected light or epi-illumination mode. In this case, the light passes through the objective before being focused on the thick or opaque sample and the reflected light from the sample is captured by the same objective lens. The properties of minerals in PPL are described as follows.[2]

7.2.1 Color

Minerals have natural or body ranging from colorless (feldspar and quartz), brownish (biotite), yellow (staurolite) and green (hornblende). White light consists of all the wavelengths between the two extremes of violet (390 nm) and red (760 nm). When white light passes through a mineral in a thin section, some wavelengths may be absorbed while others are transmitted. The color observed thus depends on the combination of wavelength transmitted. For colorless, transparent minerals such as quartz, all the wavelengths are transmitted and white light makes the mineral appear colorless. For opaque metallic minerals, all the visible light spectrum wavelengths are absorbed and the minerals will appear black. For minerals that show colors, the wavelengths of the light beam are selectively absorbed so that the resulting color of a mineral will represent a combination of the wavelengths not absorbed or transmitted.

7.2.2 Pleochroism

This property refers to the change in the color of a mineral between two extremes when oriented differently in a complete 360° rotation on the microscope's stage. The behavior arises because the minerals, for instance, iron-magnesium dominant minerals such as biotite, amphiboles and staurolite absorb light differently in different orientations.

7.2.3 Habit

This refers to the shape a mineral type shows in different rocks. It may appear euhedral when it has well-defined crystal surfaces or anhedral. Other shapes include:

  i. Prismatic—if the crystal is elongated in one direction, e.g., apatite
 ii. Acicular—if the crystal appears like a needle
iii. Fibrous—when the crystals look like fibers
 iv. Tabular or platy—when the crystals are flat and thin, e.g., biotite

7.2.4 Cleavage

There are planes of weakness in the mineral's atomic structure and most minerals are cleft along certain specific crystallographic directions that are related to such planes. Some minerals like quartz and garnet have no cleavages. Other properties of minerals include relief and alteration.

7.3   Reflected Light Microscope

Reflected light microscopy is used to study opaque materials that include geological samples (polished sections), building materials, metals and opaque particles. The light source is located above the sample and is focused on the sample. The surface imaging technique enables the determination of surface features such as cracks, luster, color and tool marks to provide clues about an unknown particle's identity. Reflected light microscopy can be coupled with interferometry to study nanometer scale surface fluctuations and with polarized and ultraviolet lights to identify minerals and phases as well as to study fluorescence, respectively.

In reflected light microscopy, the sample is mounted with epoxy resin and the surface is polished. The surface prepared is viewed with a microscope in a reflected light mode. After the image is taken with the reflected light microscope, the surface can be covered with evaporative carbon to make it conductive and suitable for scanning electron microscopy. The mineral phases present will be identified by the colors obtained from their reflections. For a Brazilian iron ore sample, the mineral phases were identified with colors as follows- hematite (colorless), magnetite (purple), goethite (light grey), quartz (dark grey) and epoxy resin background (black).

The two main types of reflecting microscopes are the simple binocular microscope and scanning electron microscope. A binocular microscope is an optical microscope with two eyepieces, unlike a monocular one, to ease viewing significantly and to reduce straining of eyes. Most microscopes in use today are binocular although there are differences in the interplay between the two eyepieces depending on the type of microscope. Other differences in binocular microscopes include the light source used, the minimum and maximum magnification possible and the availability of an image-saving method. Figure 7.1 shows the labelled diagram of a monocular microscope.[3, 4]

The advantages of a scanning electron microscope over conventional microscopes are higher resolution and very high magnification that can be up to 2,000,000

7.4   Thin Section Transmitted Light Experiment

The CETI light transmission microscope can be used to obtain thin section transmitted light photomicrograph. A mixture of Araldite resin and hardener are mixed together thoroughly in equal proportion in a square container. Thereafter, 0.5 g of the screened ore is poured into the mixture prepared and further mixing is carried out. The complete mixture is then placed on a glass slide of a rectangular shape and then left on a table for about an hour to get hardened. After hardening, a grinding wheel machine is applied to thin the sample on the glass slide, while the finishing thinning to the appropriate diameter is done on a lapping/thinning plate which is sprayed with silicon carbide. If the ore is wet, the sample is heated on a hotplate for about 5 min for drying. After the drying, Canada balsam paste is placed on the sample surface and a cover slip is used to cover the surface to preserve it. Thereafter, the prepared slide is viewed under the microscope in the light transmission mode at the magnification of ×400. The microscope is adjusted for proper viewing to obtain the best possible view to be taken with a digital camera.

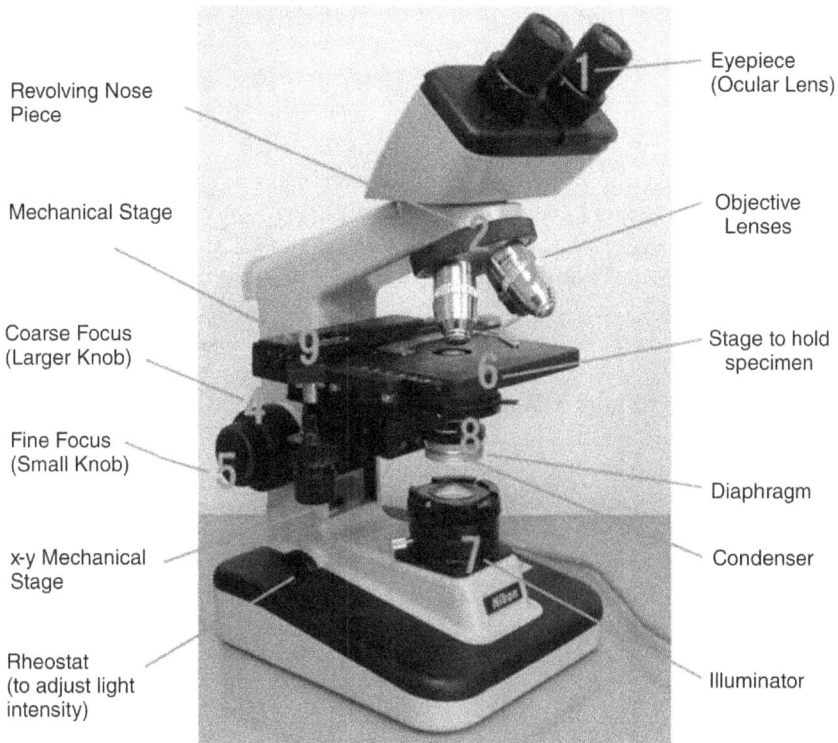

FIGURE 7.1 Components of a typical brightfield microscope.[5]

Source: **Courtesy of pinterest.com (2022)**

7.5 Ore Chemical Analyses

When a fresh ore is received in the laboratory, it is necessary to subject the ore to analysis to determine its chemical composition and thus be able to properly identify it. There are four important factors to be considered in the chemical analyses of ore minerals, namely.[6]

1. Specification of the elements to be determined in the ore
2. The chemical nature of the elements, that is, atomic number and atomic mass
3. The expected concentrations of the elements in either weight%, ppm, ppb or ppt
4. Identification of the minerals or mineral phases with which the elements are associated

Aside from the mineralogical analysis, there are three categories of analysis, namely: wet classical analysis, powder or wet spectroscopy and in-situ spectroscopy that can be used. In wet classical analysis, the ore is dissolved in an acid or other corrosive solutions and the resulting solution is analyzed. In spectroscopy tests, an energy source is used to bombard an ore to generate an electromagnetic signal that is then detected and analyzed.

The wet chemistry method may be gravimetric, volumetric or calorimetric. The gravimetric method determines the chemical composition of a material using weight measurements, while the volumetric methods use volume measurements. On the other hand, calorimetric methods measure changes in state variables of a material in a calorimeter to derive heat transfer arising from a change of state during physical changes, chemical reactions or phase transition. These wet methods require relatively "pure" powder samples and this poses a problem in the application of these methods.

The many techniques of spectroscopy depend on the following principles:

1. An ore sample, with its aggregate of minerals, is bombarded with a high kinetic energy particle such as an electron to cause the electrons and/or neutrons in it to become excited to higher energy levels in it.
2. The excited electrons/neutrons will afterwards relax to their ground lower energy levels and release energy characteristic of each element.
3. Detectors are arranged to collect one of the following:
   a. Specific energies characteristic of each element
   b. Specific wavelengths characteristic of each element
   c. Specific frequencies characteristic of each element

The spectroscopic techniques include the following:

7.5.1   Inductively Coupled Plasma and Atomic Absorption Spectroscopy (ICP & AAS)

The basic features of these techniques are:

1. The samples are prepared as solutions using the wet chemistry dissolution approach.
2. The sample solution is sprayed into a chamber where it is heated to high temperatures of up to 6600°C to form an aerosol.
3. In AAS, the aerosol is passed through a controlled flame which absorbs the lights of specific elements emitted by particular elements in the aerosol.
4. The AAS is equipped with a monochromator linked to a detector to determine and quantify the absorption and concentrations.
5. In ICP, a gas such as argon is used to lead the sample solution into the chamber where the argon gas and the sample solution are heated to higher temperatures than in AAS for the sample to produce lights with wavelengths characteristic of its elements which are also detected and quantified as in AAS. The higher temperatures in the ICP make it to provide higher detection sensitivity and this is an advantage since many metals have sensitive ionic emission lines and ICP is used to detect metals and several non-metals in liquid samples at very low concentrations.

7.5.2 X-ray Fluorescence and X-ray Diffraction Techniques

The basic features of these techniques are:

1. X-ray fluorescence (XRF) technique determines the elemental composition while the X-ray diffraction technique (XRD) determines the compound or mineral phase composition.
2. A high-velocity electron is used to bombard a metal target to produce a monochromatic X-ray beam that hit the sample.
3. The hitting of the sample causes the ejection of the inner shell electrons from atoms of each constituent element which are then replaced by outer shell electrons. This phenomenon is called fluorescence.
4. This resulted in the emission of photons of energies characteristic of each element present and are identified and quantified by comparison with standards. XRF can analyse most elements with an atomic mass higher than oxygen down to ppm level.
5. The XRD technique is similar to the XRF technique but it depends on the diffraction of the X-ray beam incident at a known range of angles by the ore grains or crystals in the ore based on the Bragg law. This enables the determination of mineral compounds and phases from diffraction peaks at characteristic Bragg angles on the XRD chart.
6. Although the X-ray beam is incident at the sample target at a set of angles, the goniometer, which is the distance between the X-ray source and the target sample, is kept constant.

Handheld X-ray fluorescence analyzers are widely used in geological exploration. They provide real-time, on-the-spot analysis for gold pathfinder elements such as As, Cu, Pb, Zn, Ag, Hg, Co, Ni, Sb, Te and Se and thus make exploration faster.

7.5.3 Electron Probe Microanalysis (EPMA)

The basic features of this technique are:

1. An electron gun is used to generate a beam of electrons that is set to focus on the sample using a system of lenses and apertures.
2. The beam of electrons interacts with the material of the specimen to produce the excitation of the sample electrons which on relaxing to their ground states will fluoresce and produce X-rays.
3. The X-rays produced have wavelengths and photon energies characteristic of elements in the sample and can therefore be used to identify them while their intensities are used to determine corresponding concentrations in relation to the appropriate standard.
4. The Energy Dispersive Spectrometer (EDS) uses the magnitude of the energy of emitted X-ray photons, while the Wavelength Dispersive Spectrometer (WDS) uses crystals to detect particular wavelengths in the emitted photons.

In Electron Microprobe Analysis, selected grains in the polished section are focused on using a microprobe such as an SX50 microprobe analyzer to obtain the chemical composition of the phases in an ore. The instrument is typically operated at 20 kV and 30 nA probe current.

7.5.4   Scanning Electron Microscope (SEM)

The basic features of this technique are:

1. The interaction of electron beams from the SEM gun with the sample produces backscattered, secondary and Auger electrons as in reflection mode.
2. The sample images are obtained from backscattered and secondary electrons. The secondary electrons yield images that provide information about the topography of the sample while the backscattered electrons indicate contrasts in the sample surface based on the average atomic number.
3. SEMs are commonly equipped with EDS.

Figure 7.2 presents the schematic diagram of a scanning electron microscope.[7]

7.5.5   Transmission Electron Microscope

Transmission electron microscopy (TEM) is a microscopy technique in which a beam of electrons is transmitted through a specimen to form an image. TEMs are

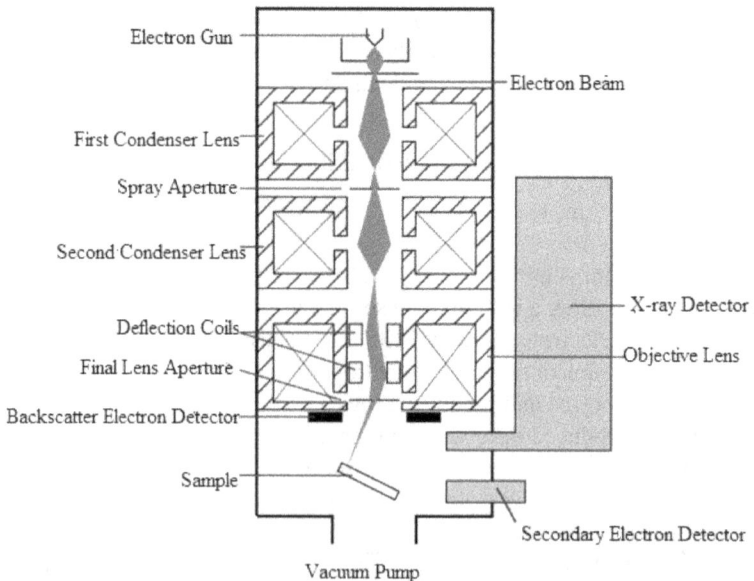

FIGURE 7.2   Schematic diagram of a scanning electron microscope.

Source: **Courtesy of Wikimedia (2022a)**

capable of imaging at a significantly higher resolution than light microscopes, owing to the smaller de Broglie wavelength of electrons. The main difference between SEM and TEM is that SEM creates an image by detecting reflected or knocked-off electrons while TEM uses transmitted electrons (electrons which pass through the sample) to create an image. The image contrast in TEM is produced based on the density variation on the specimen arising from differences in atomic number. The electron absorption is directly proportional to average density and so the brightness observed from areas decreases with increasing density. Figure 7.3 presents the labelled diagram of a TEM.[8]

7.5.6   Mass Spectrometry

The basic features of this technique are:

1. Mass spectrometry uses ion beams and is very suitable for determining the concentrations of very light elements such as H, Li and Be because the ion beam is relatively low energy so that light elements or substances in low concentrations can be quantified.
2. There are several types of mass spectrometers such as Secondary Ion Mass Spectrometer (SIMS), Thermal Ionization Mass Spectrometer (TIMS) and Multicollector Mass Spectrometer.
3. In this technique, an ion beam is generated from a plasmatron or an alkali metal and it is focused on the metal sample to erode and sputter the target area. The sputtered ions are then carried by a counter gas into the spectrometer. The technique involves the following steps:
   a.   The atoms in the sample are ionized to form cations
   b.   The ions formed are accelerated so that they all have the same kinetic energy
   c.   The cations are then deflected by a magnetic field based on their atomic number and the number of electrons ejected such that the deflection increases with decreasing atomic number and increasing positive charge number
   d.   The beam of ions deflected is then electrically detected

Figure 7.4 shows the schematic diagram of a mass spectrometer.[9]

7.6   Visible and Infrared Spectroscopy

The basic features of this method are:

1. In these methods, the concentrations of substances are determined based on the absorption or non-absorption of radiations within certain light wavelength ranges of the electromagnetic spectrum. These include ultraviolet, mid-infrared, infrared and far infrared light produced by tungsten-halogen or deuterium light sources.
2. The Beer-Lambert law is then applied. The selected light type is passed through a series of dilute solutions with known concentrations in ppm

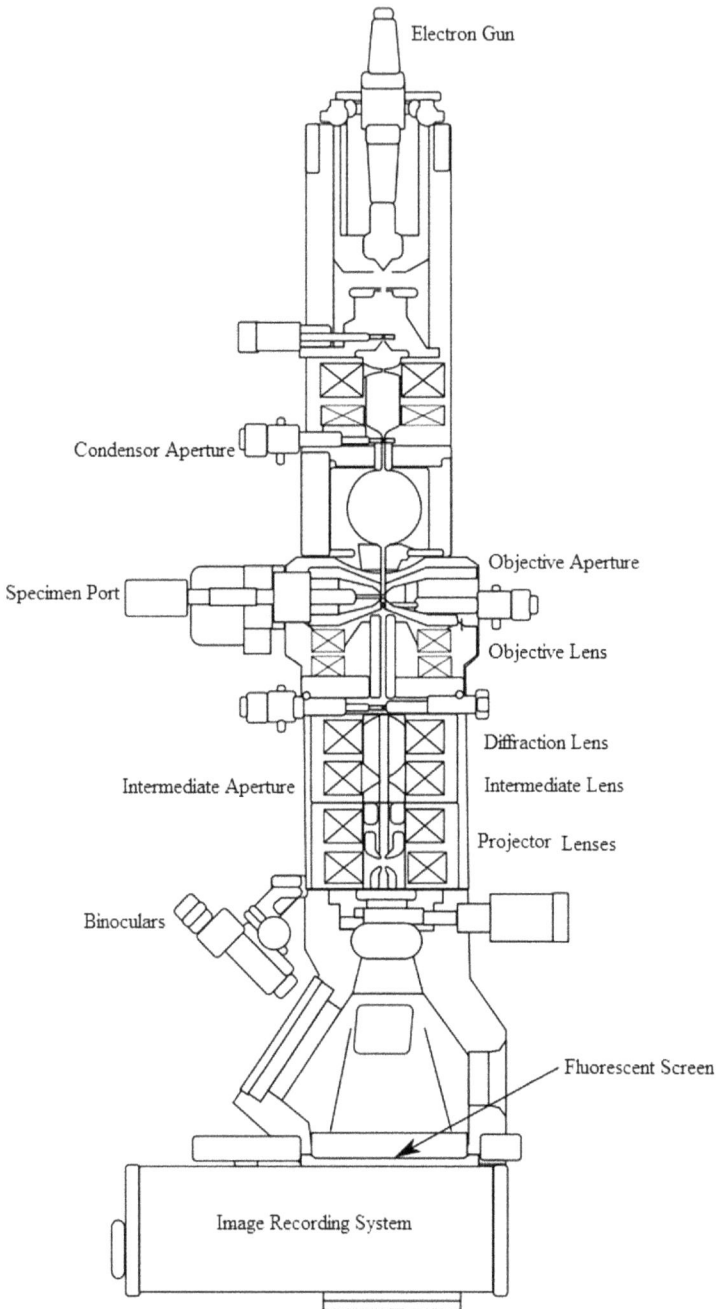

FIGURE 7.3   Transmission electron microscope.

Source: **Courtesy of Wikimedia (2022b)**

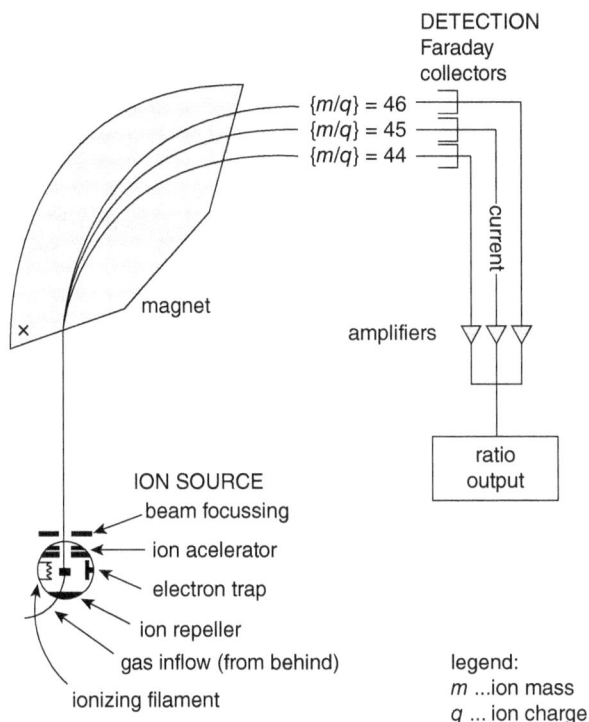

FIGURE 7.4   A mass spectrometer.

Source: **Courtesy of Wikimedia (2022c)**

containing the analyte to be determined in the sample and the amount of the incident light absorbed is measured. A plot of absorbance against concentration is done and the slope is determined. Then, the absorbance of the sample with an unknown analyte concentration is also determined. The concentration of the analyte in the sample solution is then determined from the standard curve plotted, or by calculation.

In addition to the aforementioned techniques, there are techniques based on the use of nuclear reactors. These include:

1. Neutron activation analysis
2. Photon-induced emission spectroscopy using gamma-ray and X-rays

References

1. Wills, B.A., and Napier-Munn, T.J. (2006): *Mineral Processing Technology*, 7th Edition. Amsterdam: Elsevier Science and Technology Books.

2. Munyao, N.C. (2002): *Lecture Series: SGL 201—Principles of Mineralogy* (https://profiles. uonbi.ac.ke/cnyamai/classes/sgl-201-principles-mineralogy, Accessed 26th June, 2020).
3. Gomes, O.F.M., and Paciornik, S. (2008): *Iron Ore Quantitative Characterisation Through Reflected Light-Scanning Electron Co-Site Microscopy.* Ninth International Congress for Applied Mineralogy Brisbane, QLD, 8–10 September.
4. Microtrace (www.microtrace.com/technique/fluorescence-microscopy/, Accessed 27th June, 2020).
5. Pinterest (2022): *Components of a Typical Brightfield Microscope (Courtesy of Pinterest, 2022)* (https://i.pinimg.com/originals/d8/45/40/d8454047a7d1edd59850608f6eced000. jpg, Accessed 4th May, 2022).
6. University of Massachusetts (UM): Chemical Analysis of Minerals: Quantitative Methodology in Mineralogy and Mineral Physics (www.geo.umass.edu/courses/geo311/ min%20chem%20analysis%20lecture.pdf, Accessed 26th June, 2020).
7. Wikimedia (2022a): *Scanning Electron Microscope Wiki* (https://upload.wikimedia.org/wikipedia/commons/thumb/0/0d/Schema_MEB_%28en%29.svg/1024px-Schema_MEB_%28en%29.svg.png, 240422) (Courtesy Wikimedia) Author to attribute By File:Schema MEB (it).svg: User:Steff, modified by User:ARTEderivative work MarcoTolo—File:Schema MEB (it).svg, CC BY-SA 3.0, https://commons.wikimedia. org/w/index.php?curid=9643934, Accessed 22nd April, 2022).
8. Wikimedia (2022b): *Transmission Electron Microscopy* (https://upload.wikimedia.org/wikipedia/commons/thumb/2/25/Scheme_TEM_en.svg/800px-Scheme_TEM_en.svg. png, Accessed 10th May, 2022) (By Gringer (talk)—Commons: Scheme TEM en.png, CC BY-SA 3.0 (https://commons.wikimedia.org/w/index.php?curid=5624170, 1005 en.wikipedia.org/wiki/Transmission_electron_microscopy).
9. Wikimedia (2022c): *Mass Spectrometer* (https://upload.wikimedia.org/wikipedia/commons/0/0d/Mass_Spectrometer_Schematic.svg, 0405).

8 Ore Screening

8.1 Introduction

Industrial screening uses screens which are flat metal surfaces with apertures or holes usually of uniform dimension for particle size separations. Screening has to do with size separations and can be categorized as industrial screening and industrial classification depending on the size fractions involved.

Industrial screening or sizing refers to size separations of crushed ores with sizes in the range of ≤300 mm to ≥40 μm and the separation efficiency rapidly decreases as the particle fineness increases. Industrial classification refers to size separations at finer particle sizes. Between 250 and 40 μm sizes, size separations are preferably done with classifiers although there are screens that can also efficiently do them.

There are three types of size control, that is, scalping, circuit sizing and product sizing. In scalping, the undersize is not allowed to enter the next reduction stage; while in circuit sizing, the oversize is prevented from entering the next reduction stage but it is returned to the previous stage for re-crushing. In product sizing, the output from the crusher is subjected to screening to produce products of different size ranges. Screening can be done dry or wet:

Dry: 300 mm down to ≥5 mm
Wet: <5 mm to ≥250 μm
But < 250 μm to ≥40 μm are preferably treated with industrial classifiers

Screening is the most common method of separating ore particles based on size. For industrial screenings of ores, the following basic facts are important:

  a. For pre-screening and sizing operations, screens with 20-degree inclination are used.
  b. A decrease in this inclination will slow down the sliding motion across the flat surface and thus increases screening efficiency but decreases screen capacity.
  c. For particles coarser than 12 mm, screening ore dry is preferred, while for particles finer than 12 mm, wet screening is done using water spraying at low pressure and flowing with a velocity of 0.8–1.4 m³/t/h.[1, 2]

8.2 Main Purposes of Industrial Screening

The main purposes of industrial screening are:

  i. Sizing or classifying is done to separate particles based on size difference with the usual aim of providing a unit process downstream with particles in the size range suited for that unit operation. For example, froth flotation

DOI: 10.1201/9781003323433-8

requires feeds that are <300 μm in size, while gravity concentrations require ore feeds with sizes >300 μm.

ii. It is used for scalping.

The coarsest size fractions in the ore feed are scalped, that is, they are removed from the succeeding crusher's inlet so that they can be returned for further crushing in the preceding stage or taken away from the process entirely.

iii. It is used for grading.

Combination of the size fractions as required based on specifications. Grading deals with the supply of products that requires a combination of standard-size fractions. There are size specifications in quarry and iron ore supplies.

iv. It is used for media recovery.

This is done in the dense medium separation of coal and other ore minerals. Dense media such as magnetite or ferrosilicon are washed to be free from the ore minerals.

v. It is used for dewatering, that is, removal of water from a wet sand slurry.

vi. It is used for de-sliming or de-dusting, that is, the removal of slime or dust, which are fine materials from a wet or dry ore feed with particle sizes generally smaller than 0.5 mm.

vii. It is used for taking away trash such as weed fibres from a slurry stream.

Figure 8.1 shows an Eriez vibratory screen feeder with variable vibration intensity for improved screening.[3]

In screening, two basic processes, that is, stratification and probability of particle separation are involved. Stratification is the process of escaping of large-size particles to the surface of the bed of material on the vibrating screen, while the small particles move through the voids in the bed of material to form the bottom layer. The

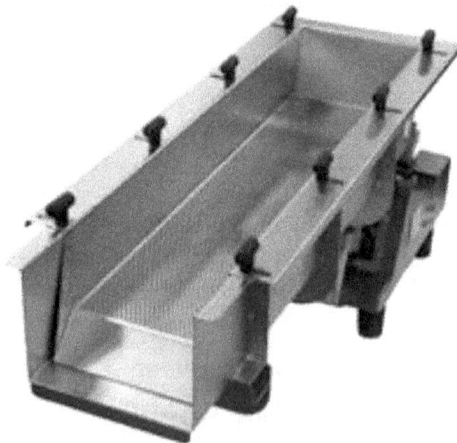

FIGURE 8.1   Eriez vibratory screen feeder.

Source: Courtesy of Eriez (2022)

probability in this case is a measure of the chance of a particle reporting to a screen aperture and passing through it if it has a size that is smaller than the aperture size. The two processes are due to the screen vibration. Each vibration lifts the particles up, making them stratify, while the surface slanting makes the particle cascade down the slope, introducing a chance for it to pass through the screen.

The screen's width is selected based on a given feed rate so that the bed depth will be under control and hence particle stratification will be optimized. When the particles delivered to the screen weigh less than 1.29 kg/m^3, then the depth of the bed at the discharge point should not exceed three times the aperture size, but if the weight is higher; the bed depth is restricted to less than 4 times the aperture size. After stratification, the probability of a mineral particle passing a screen aperture after one trial decreases with the magnitude of the particle size.[2]

8.3   Screen Performance

The particles to be presented to a screen to be screened will either pass through the screen or be retained on the screen depending on whether the size of the particles is larger or smaller than the screen's nominal aperture or its governing dimension. The efficiency of screens refers to the percentage of the misplaced particles in the desired undersize product fraction as oversize or vice versa.

Consider a screen fed with an ore such that:

F = feeding rate unto the screen in tph
f = fraction of the material in the ore feed with sizes above the screen cut point
　　size
c = fraction of the material above the screen cut point size in the overflow
u = fraction of the material above the screen cut point size in the underflow
　　such that f, c and u are obtained by sieving representative samples of the
　　feed, the overflow and underflow on a screen with the same aperture size as
　　the industrial screen
C = coarse product that overflows from the screen in tph
U = fine product that passes through the screen in tph

$$\text{Then } F = C + U \tag{8.1}$$

Mass balancing on the oversize material gives

$$Ff = Cc + Uu \tag{8.2}$$

and the mass balance of the undersize material is

$$F(1-f) = C(1-c) + U(1-u) \tag{8.3}$$

From the equations, we can derive that

$$\frac{C}{F} = \frac{f-u}{c-u} \text{ and } \frac{U}{F} = \frac{c-f}{c-u} \tag{8.4}$$

And the combined efficiency or overall efficiency can be derived as

$$E = \frac{(c-f)}{c(1-f)}$$

(8.5)

8.4 Efficiency or Partition Curve or Tromp Curve

Partition or tromp curves are used in the industries such as coal and mineral processing to determine the efficiency of a process. The curve for size distribution, is obtained when the partition coefficient (PC), which is the percentage of the feed which reports to the oversize product is plotted against the geometric mean size (GMS) on a logarithmic scale for the different size fractions of the ore. It shows how the ore particles of different size fractions in a given ore respond to the screening process in relation to the cut or separation size. The cut point is the geometric mean size that corresponds to the 50% probability, that is, the size at which 50% of the ore particles report to the undersize and 50% report to the oversize. It can be seen from the curve, Figure 8.2, that as the specific gravity, SG, increases towards the cut SG the PC increases for separation based on density.

The curve can also be constructed for the size fractions based on other characteristics such as relative density, magnetic susceptibility or charge conductivity in relation to their response to ore separation processing. The geometric mean size for

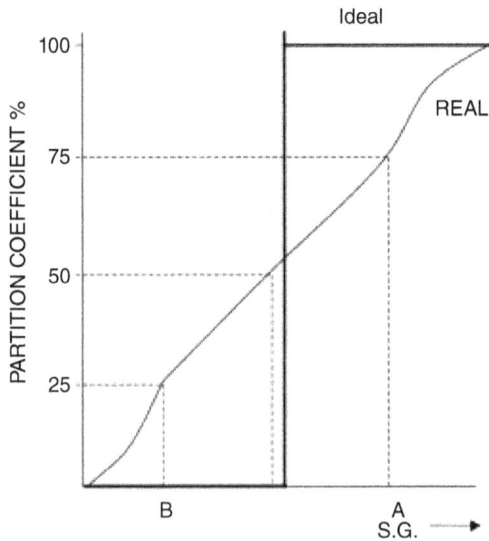

FIGURE 8.2 A Tromp curve.

Source: **Courtesy of 911 Metallurgist (2022a)**

particles in the range, say −75 + 63 mm, which implies that the particles pass the 75 mm screen but are retained on the 63 mm screen is $\sqrt{75 \times 63} = \sqrt{4725} = 68.74$ mm. Figure 8.2 shows a Tromp curve.[4]

The separation size or cut point size is the point on the abscissa at which a particle on the screen has an equal probability of reporting as undersized or oversize products and it is obtained at 50% probability on the ordinate and hence it has the symbol D50. The steepness of the curve, For the tromp curve relating to density separation, Ecart Probable, Ep, indicates the efficiency or the sharpness of the separation and is given by (D75-D25)/2 where D75 is the relative density when the partition coefficient is 75% and D25 is the relative density when the partition coefficient is 25%.

The Tromp curve can be used to evaluate the efficiency of mineral processing separators. The performance of separators can be quantified by using partition analysis. A partition curve indicates the probability of a given particle having particular characteristics to report to a given product stream. It can be applied to all types of separators such as those based on particle size, density, magnetic susceptibility, floatability and so on. Consider, for instance, a particle size-based separation at say 0.40 mm cut size. The cut size or cut point or D50 is the size such that 50% of the particles are larger and 50% of the particles are smaller in relation to the midpoint particle size. If 100% of the particles that are larger than 0.4 mm report to the oversize stream and 100% of particles that are smaller than 0.40 mm report to the undersized stream, then a perfect separation is obtained. However, if the separation is not perfect, the misplacement of particles to the wrong streams will occur. This means that some >0.40 mm particles will report to the undersize stream and vice versa.

The laboratory dense medium separation uses heavy organic liquids and is considered ideal as no particle misplacement is expected. However, in industrial dense medium separation, heavy media such as suspension of magnetite in water are used and the inefficiencies in the separation process become obvious, that is, particles denser than the separating density are misplaced into the floats and vice versa. The inefficiency arises when an ore particle has the same density or a density close to that of the separating medium and so has a chance of reporting to either the float or the sink. This condition can be represented by a partition curve. The lower the Ep, the more idealized the separation becomes since an ideal separation has an Ep of zero. The sink and float tests provide information about the applicability of pre-concentration to a particular ore. If the feed to a dense medium separation process consists of 120 tons/h of 1.50–1.56 tons/m^3 of materials and the product concentrate contains 85 tons/m^3 of the same density range material, then the partition coefficient is 85/120, which equals 0.71% or 71% partition coefficient with respect to the product or 29% with respect to the reject (911 Metallurgists, 2020a).[4]

8.5   Screen Types

There are shaking, reciprocating and vibrating screens. The vibrating screen is presently the dominant screen type while the shaking and reciprocating screens are now obsolete. Vibrating screens have a rectangular screen surface with feed and oversize discharge points at opposite ends. They perform size separation from 300 mm in size to 40 μm. They have applications in scalping, dewatering, wet screening and washing.

In multiple deck systems, the feed is introduced to the top coarsest screen with the undersize falling through the lower screen decks and thus producing a range of sized fractions from a single screen input. Inclined vibrating or circular motion screens are widely used for sizing operations. A vertical circular vibration or elliptical vibration is induced mechanically by the rotation of unbalanced weights or flywheels attached usually to a single drive shaft. The amplitude of the throw can be adjusted by adding or removing weight elements bolted to the flywheels.[1]

8.6 Grizzly Screens

A grizzly screen is an inclined surface with holes through which large-size ore materials are passed. They are made of parallel wear-resistant manganese steel bars or rails. The gap between the grizzly bars or rails is usually not larger than 300 mm but greater than 50 mm with the feed ore top size that can be as large as 1000 mm. Vibrating grizzlies are equipped with a circular throw mechanism and are usually inclined at an angle of 20°. The largest grizzlies can have capacities exceeding 5000 tph. The most important use of grizzly in mineral processing is for size control of ore feed into primary and secondary crushers. If, for instance, a crusher has a 150 mm screen setting, then the input feed can be passed over a preceding grizzly with a 150 mm gap between the bars in order to reduce the load on the crusher. Figure 8.3 shows a two-deck rectangular screen.[5]

8.7 Industrial Screening

Adeleke et al. (2013)[6] reported the improvement obtained in the crushing capacity of an aluminosilicate crushing plant by the introduction of a scalping screen for the jaw crusher discharge prior to delivery into the secondary gyratory crusher. The crushing plant design is shown in Figure 8.4.

FIGURE 8.3   A two-deck rectangular screen.

Source: **Courtesy of 911 Metallurgist (2022b)**

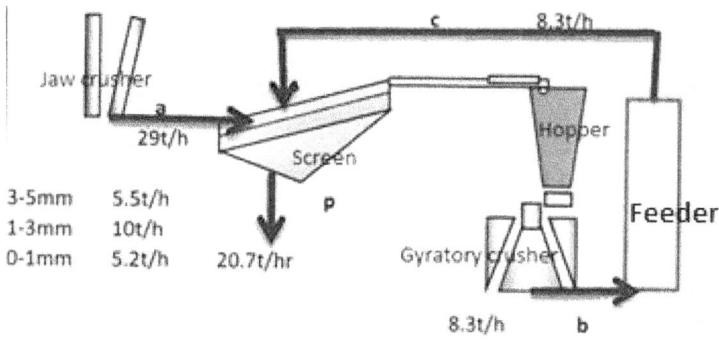

FIGURE 8.4 Introduction of a scalping screen (Adeleke et al., 2015).

References

1. Wills, B.A., and Napier-Munn, T.J. (2006): *Mineral Processing Technology*, 7th Edition. Amsterdam: Elsevier Science and Technology Books.
2. University of Alaska, Fairbanks (UAF) (2020): *Mining Mill Operator Training: Lesson 2: Classifying Cyclones* (https://millops.community.uaf.edu/amit-145/, Accessed 12th August, 2020).
3. Eriez (2022): *Vibratory Screeners* (www.eriez.eu/EU/EN/Products/Vibratory-Feeders-and-Conveyors/Vibratory-Screeners.htm, Accessed 26th May, 2022).
4. 911 Metallurgist (2022a): *Tromp Curve* (https://z4y6y3m2.rocketcdn.me/blog/wp-content/uploads/2015/06/Tromp-Curve-Example-of-Partition-Curve.png, Accessed 22nd April, 2022).
5. 911 Metallurgist (2022b): *A Two Deck Rectangular Screen* (www.bing.com/th?id=OIP. FVSuwW5qTwEr8JRS2b1RgwHaGF&w=103&h=88&c=8&rs=1&qlt=90&o=6&pid=3 .1&rm=2, Accessed 26th May, 2022).
6. Adeleke, A.A., Popoola, A.P.I., and Mekgwe, T. (2013): The Improvement of the Efficiency of an Aluminosilicate Crushing Circuit. *Journal of Mining and Metallurgy*, 49A(1), 31–36.

9 Ore Classification

9.1 Introduction

Ore classification deals with size separation using wet classifiers at fine particle sizes below 250 μm where screening becomes in-efficient. It involves the separation of a mixture of mineral particles in a slurry based on the terminal velocity each particle attains under the action of gravity.

9.2 The Principles of Classification

When a particle is introduced into a viscous carrier fluid in a sorting column with an upward velocity of V, it is going to begin a downward motion due to the gravitational force (Fg) with its acceleration decreasing because of fluid resistance. The fluid's resistance to the downward motion consists of a viscous resistance due to the viscosity of the fluid (Fv) and a resistance (Ft) due to the carrier fluid's displacement by the particle and this may be laminar or turbulent depending on the particle velocity. At a point, an equilibrium will be reached between Fg and (Fv + Ft) and the particle will attain its terminal velocity and subsequently descend at a uniform rate. The particle in question may be small or large and may be light or heavy depending on its density, size and shape.[1]

The total resistance to the particle's downward motion is given by

$$R = Fv + Ft \tag{9.1}$$

and by extension, the terminal velocity (Vt) that the particle will attain depends on its density, size and shape.

If Vt < V a particle will float as an overflow
Vt > V the particle will sink as an underflow

The coming down or descent of a particle suspended in water is also called settling or dynamic settling. A settling particle may experience a free or hindered settling depending on a number of factors.

Classifiers carry out size control using the motion of particles that depend on size, shape and specific gravity.

9.3 Settling of Ore Particles in Carrier Fluids

9.3.1 Conditions for Particles Settling

The conditions for free settling of particles in a fluid are:

a. The particle's content of the slurry is such that the slurry density or pulp density expressed as % solids are less than 15%

DOI: 10.1201/9781003323433-9

b. At such a pulp or slurry density, particle crowding will be negligible
c. The governing density of the slurry will be the density of the carrier fluid and not that of the slurry or pulp

On the other hand, the conditions for hindered settlings are:

a. The particle content at the slurry is such that the pulp or slurry density, expressed as % solids greater than 15%
b. In such a case, the pulp density is such the particle crowding becomes appreciable
c. For such a slurry, the governing density is the slurry density and not the carrier fluid density

9.3.2 The Forces Acting on a Particle

Figure 9.1 shows a spherical particle having a diameter d and density Ds falling in a viscous fluid of density Df under gravity in a free settling condition. The particle falls under the downward-acting gravity force Fg, the upward-acting viscous resistance Fv and the buoyancy resistance due to fluid displacement (Ft). The particle's equation of motion is given by:

$$mg - m'g - D = \frac{mdy}{dt} \tag{9.2}$$

where y = the velocity of the particle in the y-direction
D = total upward acting force = Fv + Ft
m = the mass of the ore particle
m' = the mass of the fluid volume displaced
$\frac{dy}{dt}$ = acceleration of the particle under the forces shown
When the particle attains its terminal velocity, $\frac{dy}{dt}$ becomes zero
Therefore $mg - m'g - D = 0$

$$D = mg - m'g = (m - m')g$$
$$D = \left(\frac{\pi}{6}\right)gd^3\left(D_s - D_f\right) \tag{9.3}$$

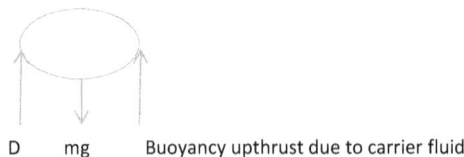

D   mg   Buoyancy upthrust due to carrier fluid

FIGURE 9.1   Forces acting on a particle in a carrier fluid.

Based on the assumption that the drag force D originates from the viscous resistance Fv only, Stokes derived for a particle with a diameter less than about 50 µm terminal velocity v_t as:

$$v_t = \frac{gd^2 (Ds - Df)}{18\eta} \tag{9.4}$$

where η is the viscosity.

Newton in his own analysis assumed that the drag force D is due entirely to the fluid's turbulent resistance Ft and derived the equation for the terminal velocity of a particle with a diameter larger than about 0.50 cm as

$$v_t = \left[\frac{3gd (D_s - D_f)}{D_f} \right]^{1/2} \tag{9.5}$$

9.4 Conditions for Equal Settling Rate of Particles

For two different ore particles A and B with densities D_a and D_b and diameters da and db, respectively, falling in a fluid of density D_f to fall or settle at exactly the same settling rate or terminal velocity, the following relationships apply.

9.4.1 Free Settling Conditions

From Stokes's law, for two particles with diameters less than 50 µm to have the same settling rate:

$$\frac{d_a}{d_b} = \left(\frac{D_b - D_f}{D_a - D_f} \right)^{1/2} \tag{9.6}$$

This is known as the free settling ratio, that is, the ratio of particle size of particles A and B required for the two mineral particles to fall at the same rate or to have the same terminal velocity. For larger particles with a diameter larger than 0.5 cm, Newton's law gives the condition for an equal settling rate as:

$$\frac{d_a}{d_b} = \frac{D_b - D_f}{D_a - D_f}$$

In general, for particles A and B to have the same settling rate:

$$\frac{d_a}{d_b} = \left(\frac{D_b - D_f}{D_a - D_f} \right)^n \tag{9.7}$$

For small particles that obey Stokes' law, with a diameter <50 μm n = ½
  For large particles that obey Newton's law, with a diameter >0.5 cm n = 1
  For intermediate particles, that is, particles with diameters between 50 μm and 0.5 cm n = 0.5 to 1

9.4.2 Hindered Settling Conditions

The terminal velocity of a particle settling under hindered settling conditions was derived by Newton as:

$$V = k\left[d\left(D_s - D_p\right)\right]^{\frac{1}{2}}$$ (9.8)

where D_p is the pulp or slurry density

For two particles A and B to have the same settling rate under the hindered settling condition, the hindered settling ratio is given by:

$$\frac{d_a}{d_b} = \frac{D_b - D_p}{D_a - D_p}$$ (9.9)

The free and hindered settling ratios enable the mineral processing engineer to know the diameter ratios required under both free and hindered settling conditions for particles of light and heavy minerals to have the same terminal velocity. In practice, these ratios should guide ore crushing and grinding such that the diameter conditions are not satisfied and hence the particles will settle at different terminal velocities so that separation based on size becomes feasible between the two sets of particles. In free settling, coarse particles will move faster than fine ones, while in hindered settling, the dense particle will move faster than light ones.[1]

9.5 Worked Examples

Example 9.1
  A cassiterite ore contains the cassiterite mineral value and quartz as the main gangue mineral. The specific gravities of cassiterite and quartz are 6.98 and 2.65, respectively. Determining the conditions, in terms of diameters of the particles, for the particles of the two minerals to have the same settling rates under free and hindered settling in water and in a carrier fluid with a specific gravity of 1.9, respectively. Determine the conditions for particle sizes <50 μm, >0.5 cm and intermediate. Explain the significance of your results.

Solution

Taking c to represent cassiterite and q to represent quartz.
  For free settling conditions when particles with sizes < 50 μm are involved,
    n = 0.5

$$\frac{d_c}{d_q} = \left(\frac{D_q - D_f}{D_c - D_f}\right)^n$$

$$\frac{d_c}{d_q} = \left(\frac{D_q - D_f}{D_c - D_f}\right)^{0.5} = \left(\frac{2.65 - 1}{6.98 - 1}\right)^{0.5} = \left(\frac{1.65}{5.98}\right)^{0.5} = 0.53 \qquad (9.10)$$

$$\frac{d_c}{d_q} = 0.53 \text{ or } \frac{d_q}{d_c} = 1.89$$

This implies that $d_q = 1.89 d_c$, that is, a particle of the lighter quartz that is 1.89 times as large as a particle of the heavy cassiterite will have the same free settling rate.

For the particles that are > 0.5 cm in size n = 1

$$\frac{d_c}{d_q} = \left(\frac{D_q - D_f}{D_c - D_f}\right) = \frac{1.65}{5.98} = 0.28$$

$$\frac{d_q}{d_c} = 3.57$$

This implies that $d_q = 3.57 d_c$, that is, a particle of the lighter quartz that is 3.57 times as large as a particle of the heavy cassiterite will have the same free settling rate.

For particles with diameters between 50 μm and 0.5 cm, n can be taken as 0.75 and therefore:

$$\frac{d_c}{d_q} = \left(\frac{D_q - D_f}{D_c - D_f}\right)^{0.75} = \left(\frac{2.65 - 1}{6.98 - 1}\right)^{0.75} = \left(\frac{1.65}{5.98}\right)^{0.75} = 0.275$$

$$\frac{d_q}{d_c} = 3.62$$

For hindered settling,

$$\frac{d_c}{d_q} = \left(\frac{D_q - D_p}{D_c - D_p}\right) = \frac{2.65 - 1.9}{6.98 - 1.9} = \frac{0.75}{5.08} = 0.1476$$

$$\frac{d_q}{d_c} = 6.77$$

From the results obtained, the following deductions can be made:

a. The free settling ratio is observed to be higher for particles coarser than 50 μm and the higher the ratio, the better the separation efficiency. Free settling ratio is thus more particle size sensitive than density sensitive.
b. For a particular particle, the hindered settling ratio is always greater than the free settling ratio.

    c. The denser the pulp, that is, the higher the percentage of solids, the greater the ratio of diameters for particles with equal settling rate.

    d. It has been found that in practice, the highest hindered settling ratio that can be obtained for galena and quartz is 7.5, the specific gravity of galena. Thus, the higher the density of the heavy mineral, the higher will be the hindered settling ratio and the better will be the separation efficiency between the particles of the light and heavy minerals. Hindered settling ratio is thus more density sensitive than size sensitive.

    e. The hindered settling classifiers are therefore used to take advantage of the effects of density on the separation, while the free settling classifiers take advantage of the effect of larger particle size. However, the advantage of higher density on hindered settling classification may be reduced by the increase in viscosity at higher density.

9.6 Types of Classifiers

There are two broad classes of classifiers

    1. The horizontal current or mechanical classifiers
    2. The vertical current or hydraulic classifiers

9.6.1 Hydraulic Classifier

The hydraulic classifiers consist of several vessels serving as sorting columns with different volumes such that the volume capacity increases from the first to the last. There is also an inlet for the upwardly flowing water in each vessel at different velocities v1, v2, v3 such that v1 > v2 > v3 such that a series of spigot products are obtained with the coarsest, most dense particles in the first vessel and the product becoming finer in size and less dense in the subsequent vessels.

    The slurry is fed into the first vessel and the particles in the slurry settle at different terminal velocities in the vessel depending on size, shape and density. For any of the particles descending in vessel 1, having a terminal velocity greater than the upward velocity v1 of fluid in vessel 1, the particle will sink as an underflow while other particles with terminal velocities that are lower than v1 will flow into the vessel 2 where particles that have terminal velocities greater than v2 will sink as under-flow while particles with terminal velocities less than v2 will float into the third vessel.

    Similarly, for the third vessel, the particles descending and having velocities greater than v3 will sink as underflow while particles with terminal velocities less than v3 will float as overflow such that very fine slimes overflow the last sorting column. Figure 9.2 shows a laboratory-scale hydraulic classifier.[1, 2]

9.6.2 Hydrocyclone

Hydrocyclone is a continuously operating classifier and is one of the most widely used in the mineral industry for achieving particle size separations below 300 μm.

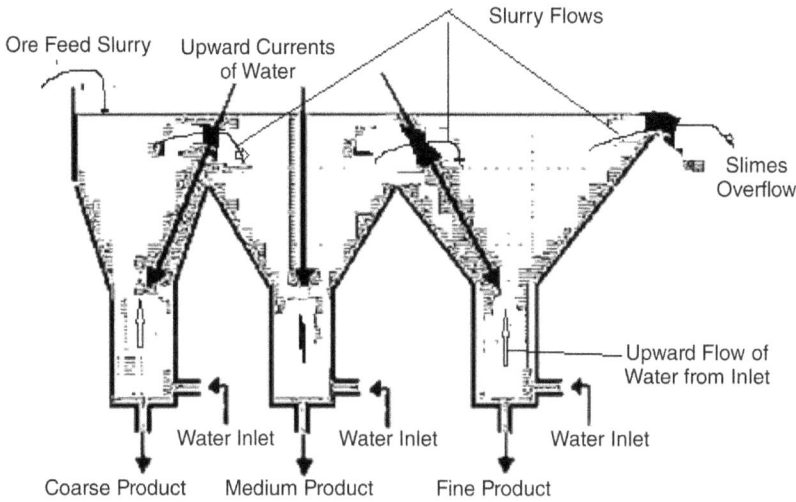

FIGURE 9.2 Laboratory-scale hydraulic classifier.

Source: **Courtesy of 911 Metallurgist (2022a), adapted**

Hydrocyclone has a cylindrical barrel with a conical section joined at the bottom. The conical section's length of the hydrocyclone has been found to have a significant effect on particle size separations. Hydrocyclone has a steel plate cover at the top on which is mounted an overflow pipe as shown.

There is a pipe extension called vortex finder linked with the over-flow pipe. The aim of the vortex finder is to prevent short-circuiting of the input slurry into the overflow pipe before it is sorted. The equipment uses input fluid pressure to produce centrifugal force and a flow pattern which can separate particles from a fluid medium based on their density, size and shape. A hydrocyclone is a classifier because it divides the particles suspended in a fluid into two fractions, ideally at a particular size, called the cut size.

The principles of operation involve the following:

   i. The slurry is pumped into the hydrocyclone under pressure through the tangential entry
  ii. The tangential feed injection induces a centrifugal force which accelerates the suspended particles settling kinetics.
 iii. The slurry and its particles are acted upon by centrifugal force and this produces the following effects:
      • A swirling action on the slurry and its particles
      • The swirling actions also generate a vortex that produces a low-pressure zone in the hydro cyclone's centre with a link into the overflow pipe and a region of high velocity near the wall of the cyclone
  iv. The centrifugal force also acts on the particles to produce different settling terminal velocities for the particles based on the size, shape and density of the particles.

v. Each particle is acted upon by two forces. The outwardly acting centrifugal force and inwardly acting drag force.

vi. Coarse particles with high settling terminal velocities (fast settling rates) will be pushed by the centrifugal force to the high-velocity region near the outer wall and will through the motion of the fluid stream move downward toward the spigot

vii. The slower-settling finer particles will be taken by the inwardly acting drag force into the low-pressure region linked to the overflow pipe where they will float into the overflow.

viii. Due to the constriction of the spigot, the volume of fluid that can report to the underflow stream is limited and thus a portion of the stream is forced to reverse its direction and move upward carrying the fine-sized particles along the low-pressure zone towards the vortex finder.[1]

Figure 9.3 shows the schematic diagram of the hydrocyclone.[3]

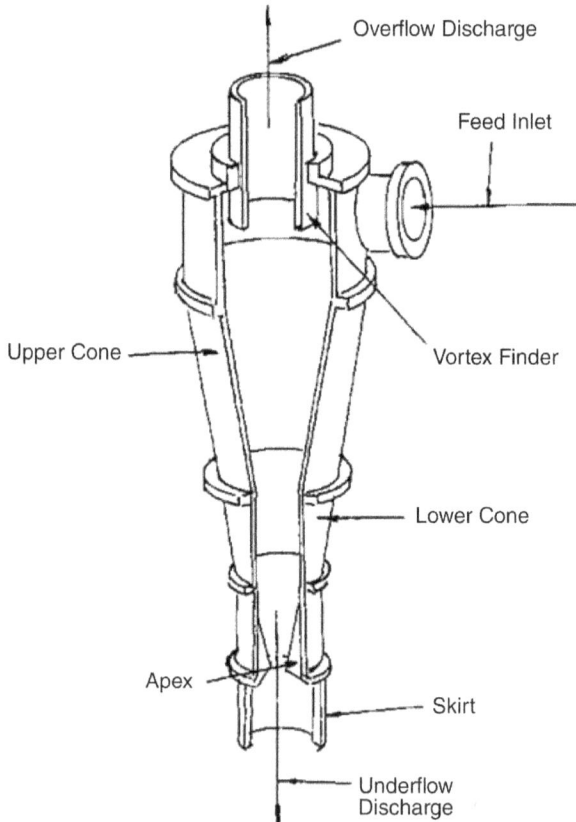

FIGURE 9.3 Schematic diagram of a hydrocyclone.

Source: **Courtesy of 911 Metallurgist (2022b)**

Hydrocyclones are used in concentrating coal run-of-mines (ROM). For instance, the South Africa South Witbank plant is a cyclone plant that treats 250 tons of ROM coal per month with fine particles passing 63 μm accounting for about 10% of the cyclone feed. The cyclone coal concentrate output typically has a calorific value of 27 MJ/kg, while the discard is further processed to obtain <82 μm underflow low-grade coals with a calorific value of 22 MJ/kg. The overflow is taken to the clarifier to recover the water for re-use, while the underflow product is dewatered for stockpiling. The dewatering screen coal overflow was found to yield a 19 MJ/kg calorific value.

Since the magnitude of the applied centrifugal force increases with a decrease in cyclone diameter, a cyclone of a particular diameter is selected based on the intended size separation. In mineral processing, a specific cut size desired is stated such that most of the solid particles in the underflow, say 95%, have sizes larger than the target cut point size and most of the particles in the overflow have sizes smaller than the stated cut point size. In design, however, the cut point size is taken as the particle size that has a 50% probability of reporting to either the overflow or underflow stream and it is called the mean particle size separation (d50). It has been established that for the small particle size separations small diameter hydro-cyclones are required.[4]

9.7   The d50 Cut Point

For cyclones that are geometrically equivalent, the cyclone's diameter determines the mean particle size and the relationship between the two parameters is stated as:

$$d_{50(c)} = D_C X \tag{9.11}$$

where D refers to the cyclone's diameter and the value of X varies for different models. For instance, X has the values 1.875 and 1.8 for Krebs-Mular-Jull (1978) and Plitt (1976) models[4], respectively.

The cyclone is expected to carry out a separation such that a particle finer than a given cut point size reports to the overflow but the design is based on the d50C parameter. In view of this, a relationship between the d50C and the overflow size has been developed. For example, to obtain 95% < 150 μm separation into the overflow, a multiplier of 0.73 is applied to obtain the required d50C. This means: $d_{50(c)} = 0.73 \times 150 = 110$ μm.

The result means that an overflow with 95% passing 150 μm is equivalent to an overflow with $d_{50(c)}$ of 110 μm. Table 9.1 shows the multiplier required for other overflow sizes.[4]

The following should also be noted regarding d50C:

a. It increases as the vortex finder diameter increases
b. It increases as the spigot diameter decreases
c. It increases as the inlet diameter increases

TABLE 9.1

Multiplier for Hydrocyclone for Different Overflow Sizes

Required overflow size (% passing)	Multiplier
98.8	0.54
95.0	0.73
90.0	0.91
80.0	1.25
70.0	1.67
60.0	2.08
50.0	2.78

9.8 Worked Example

A slurry stream with 60% solids is fed into a hydrocyclone at 1400 tph. The overflow is expected to be 80% passing 100 µm with % solids of 58% and a volumetric flow rate of 250 litre/s. If the specific gravity of the solids is 2.85, determine:

a. The specific gravity of the slurry
b. The required d50 for the specified overflow stream
c. The required cyclone diameter
d. Number of units required

Solution

a. The % solids, x, is given by

$$x = \frac{100s\left(D - w\right)}{D\left(s - w\right)}$$

where D is the slurry density, s is the density of the solids and w is the density of the water

$$x = \frac{100 \times 2.85 \times \left(D - 1\right)}{D\left(2.85 - 1\right)}$$

$$60 = \frac{285D - 285}{1.85D}$$

$$174D = 285$$

$$D = 1.64$$

b. The D50C

For 80% passing, the required multiplier with respect to 50% passing = 1.25
The specified size = 100 µm
Therefore, d50 (Actual) = 1.25 × 100 = 125 µm

c. Cyclone diameter

Since d50C = DCX

Then $125 = D_C^X$

Taking X = 1.8 and taking the log of both sides

log 125 = X log DC

$$logD_C = \frac{log125}{X} = \frac{2.097}{1.8} = 1.165$$

$$D_C = 14.62\,m$$

References

1. Wills, B.A., and Napier-Munn, T.J. (2006): *Mineral Processing Technology*, 7th Edition. Amsterdam: Elsevier Science and Technology Books.
2. 911 Metallurgist (2022a) (www.911metallurgist.com/dewatering-thickening/building-a-laboratory-scale-hydraulic-classifier/, Accessed 28th June, 2020).
3. 911 Metallurgist (2022b): *Hydrocyclone Working Principle* (https://z4y6y3m2.rock-etcdn.me/blog/wp-content/uploads/2015/08/Hydrocyclone-Operating-Principle.png, Accessed 5th May, 2022).
4. University of Alaska, Fairbanks (UAF) (2020): *Mining Mill Operator Training: Lesson 2: Classifying Cyclones* (https://millops.community.uaf.edu/amit-145/, Accessed 12th August, 2020).

10 Metallurgical Accounting Analyses

10.1 Slurry Streams Analysis

A slurry is a mixture of dry solid particles and water. The dry solid particles are transported by the water fluid medium which serves as a carrier for the particles. Grinding is mostly carried out wet and most of the post-grinding mineral processing operations are done on slurries in motion. Slurry pumps are used to transport a mixture of solids and water in several industries. The parameters of a slurry stream are as follows[1]:

1. The slurry volume

   It is important to know the volume of the slurry and the weight of the solids it contains because the slurry volume enables us to determine its residence time in each unit process while the weight of the dry solids it carries is required in metallurgical accounting calculations.
2. The mass flow rate of the slurry (M), in kg/s and the volumetric flow rate of the slurry, F in m^3/s
3. The dry solids' content mass flow rate within the slurry (M_s) and the mass flow rate of water in the slurry (M_w)
4. The density of the dry solid in the slurry (s), that is, of the particles in the slurry and can be determined according to the method described in Wills and Napier-Munn (2006) and the density of water in the slurry (w)
5. The density of the slurry or the pulp itself (D). This can be determined from standard nomograms and by calculations
6. Volumetric flow rate of water in the slurry (F_w)
7. Volumetric flow rate of dry solids in the slurry (F_s)

10.2 Determination of Volumetric Flow Rate of Slurry

For a slurry whose volumetric flow rate is not too high, the stream of the pulp can be diverted into a suitable container for a measured period of time to determine the volumetric flow rate. In such a case, the volume of slurry collected divided by the time duration of the slurry collection will give the pulp's volumetric flow rate. The slurry density is obtained by diverting small flow streams into a container of known volume. The container is then weighed to calculate the slurry density.

However, a direct reading of the pulp density can be done by filling up a density can and weighing it on a balance specially graduated.

DOI: 10.1201/9781003323433-10

10.3   The Composition of a Slurry

(1) Per cent dry solids in the slurry by weight is given by x, while % moisture or water in the slurry is given by:
Per cent water by weight = 100 − x
(2) Density of solids in the slurry (s)

A formula to determine % solids by weight, x, is derived as follows:

Assume a volume of 1 m³ of slurry with D as the slurry density, x as % solids by weight and was the density of water.

Weight of slurry = density of slurry x volume of slurry

$$= D \times 1$$
$$= D \text{ kg}$$

$$\text{Weight of solids in the slurry} = \frac{x}{100} \times D = \frac{xD}{100}$$

$$\text{Volume of solid in the slurry} = \frac{xD}{100} \div s = \frac{xD}{100s}$$

$$\text{Weight of water in the slurry} = \frac{(100-x)}{100} \times D$$

$$\text{Volume of water in the slurry} = \frac{(100-x)}{100} D \div w$$

$$= \frac{(100-x)D}{100w}$$

$$\text{Therefore } \frac{(100-x)D}{100w} + \frac{XD}{100s} = 1$$

$$x = \frac{100s(D-1000)}{D(s-1000)} = \text{per cent solids by weight}$$

$$\text{Per cent solids by volume} = \frac{xD}{s}$$

Note:
Mass flow rate of the slurry = volumetric flow rate of the slurry × slurry density

$$= F \times D \text{ kg/h} \qquad\qquad (10.1)$$

$$\text{Mass flow rate of the dry solids in the slurry} \left(M_s\right) = FD\left(\frac{x}{100}\right) = \frac{FDx}{100}$$

$$= \frac{FDs(D-1000)}{D(s-1000)}$$

where Dry solids ratio in the slurry $= \dfrac{x}{100}$ and water ratio in the slurry $= \dfrac{100-x}{100}$

$$\text{Mass flow rate of water in the slurry}\left(M_w\right) = FD \times \dfrac{\left(100-x\right)}{100}$$

$$\text{Dilution ratio}\left(W_r\right) = \dfrac{100-x}{x} \text{ and dry ratio}\left(D_r\right) = \dfrac{x}{100-x} \qquad (10.2)$$

$$\text{Volumetric flow rate of the solids in the slurry }\left(V_s\right) = \dfrac{Ms}{s}$$

$$\text{Volumetric flow rate of the water in the slurry}\left(V_w\right) = \dfrac{Mw}{w}$$

$$\text{Volumetric flow rate of the slurry }\left(M_{slurry}\right) = \dfrac{M}{D} = \dfrac{Ms}{s} + \dfrac{Mw}{w}$$

$$\text{Percent solids by volume} = \dfrac{xD}{s}$$

Given the mass flow rate of solids in the slurry as M_s

$$\text{The mass flowrate of the water in the slurry} = M_s \times \dfrac{100-x}{x}$$

Given the mass flow rate of water in the slurry as Mw

$$\text{The mass flowrate of the solids in the slurry} = M_w \times \dfrac{x}{100-x} \qquad (10.3)$$

10.4 Worked Examples

Example 10.1

A froth flotation conditioning tank of a 30 m³ volume capacity is to have 150 m³ of slurry delivered into it within 60 min.

  a. Determine what the retention time of the slurry will be in the conditioning tank
  b. What is retention time in mineral processing?

Solution

  a. Calculation of retention time
  Volume of slurry = 150 m³
  Time of slurry flow = 1 hour

$$\text{Volumetric flow rate of slurry} = \dfrac{150}{1} = 150 \text{m}^3/\text{h}$$

  This means that 150 m³ of slurry will fill a conditioning vessel in 1 hour

The retention time of the slurry in the conditioning tank will be

$$= \frac{30}{150} \times 60 = 12 \min$$

b. Retention time is the amount of time the slurry spends in the conditioning tank before its discharge after its entry into the tank

Example 10.2

Consider a slurry stream containing sphalerite solids with a density of 4100 kg/m³ is delivered into a 1-litre density can and takes 20 s to fill the can. If the density of the slurry was observed to be 3100 kg/m³, calculate:

a. The volumetric flow rate of the slurry
b. The mass flow rate of the slurry
c. The % solids by weight of sphalerite in the slurry
d. The mass flow rate of the sphalerite solids in the slurry
e. The mass flow rate of the carrier water in the slurry

Solution

a. Volumetric flow rate of the slurry $F = \dfrac{1 litre}{20\, sec} = \dfrac{1000 cm^3}{20 sec} = \dfrac{10^{-3}\ m^3}{20\, s}$

$$= 5 \times 10^{-5} m^3 / s$$

Volumetric flow rate of the slurry $= 5 \times 10^{-5} / \dfrac{1}{3600} = 0.18\, m3/h$

b. The mass flow rate of the slurry $M = FD = 0.18 \times 3100 = 558$ kg/h
c. The % solids by weight of sphalerite in the slurry

% dry solids by weight $x = \dfrac{100s(D-1000)}{D(s-1000)}$

s = density of sphalerite = 4100 kg/m³
D = 3100 kg/m³

$$x = \frac{100 \times 4100 \times (3100-1000)}{3100(4100-1000)} = \frac{816000000}{9610000} = 89.59\%$$

d. The mass flow rate of the sphalerite solids in the slurry
Mass flow rate of the slurry = 558 kg/h

The dry ratio $(D_r) = \dfrac{x}{100} = \dfrac{89.59}{100} = 0.8959$

Mass flow rate of the sphalerite solids in the slurry (Ms) = M × Dr = 558 × 0.8959 = 499.91 kg/h
Alternatively,

$$M_s = \frac{Fs(D-1000)}{(s-1000)} = \frac{0.18 \times 4100 \times (3100-1000)}{(4100-1000)} = \frac{1549800}{3100} = 499.93 \text{ kg/h}$$

e. The mass flow rate of the carrier water in the slurry

$$\text{The wet ratio}\left(W_r\right) = \frac{100-x}{100} = \frac{100-89.59}{100} = \frac{10.41}{100} = 0.1041$$

Mass flow rate of the carrier water in the slurry (Mw) = M × Wr = 558 × 0.1041 = 58.09 kg/h

Example 10.3

A pump received ore supplies from three slurry streams. The volumetric flow rate of one stream is 7.5 m³/h and it contains 42% solids by weight. The second stream has a volumetric flow rate of 3.7 m³/h and contains 52% solids by weight, while the third stream has a volumetric flow rate of 5.3 m³/h and 44% solids. If the density of solids in the slurry is 3100 kg/m³.

Calculate:

a. The slurry densities of the 3 streams
b. The mass flow rate solids in the slurry streams
c. The total mass of the dry solids pumped

Solution

For slurry 1,

$$\text{the slurry/pulp density} = D = \frac{1000 \times 100 s}{s(100-x)+1000x} = \frac{1000 \times 100 \times 3100}{3100 \times (100-42)+1000 \times 42}$$

$$= 1397.66 \text{ kg/m}^3$$

Mass flow rate of solids in slurry stream 1

$$= \frac{Fs(D-1000)}{(s-1000)} = \frac{7.5 \times 3100 \times (D-1000)}{(3100-1000)} = \frac{7.5 \times 3100 \times (1397.66-1000)}{(3100-1000)}$$

$$= \frac{9245595}{2100} = 4402.66 \frac{\text{kg}}{\text{h}} = 4.402 \text{ t/h}$$

For slurry 2,

$$\text{the slurry/pulp density} = D = \frac{1000 \times 100 s}{s(100-x)+1000x} = \frac{1000 \times 100 \times 3100}{3100 \times (100-52)+1000 \times 52}$$

$$= 1543.83 \text{ kg / } m^3$$

Mass flow rate of solids in slurry stream 2

$$= \frac{Fs(D-1000)}{(s-1000)} = \frac{3.7 \times 3100 \times (1543.83-1000)}{(3100-1000)} = 2970.35 \frac{\text{kg}}{\text{h}} = 2.97 \text{ t/h}$$

For slurry 3,

the slurry/pulp density $= D = \dfrac{1000 \times 100s}{s(100-x)+1000x} = \dfrac{1000 \times 100 \times 3100}{3100 \times (100-44)+1000 \times 44}$

$= \dfrac{310000000}{217600} = 1424.63 \, \text{kg/m}^3$

Mass flow rate of solids in slurry stream 1

$= \dfrac{Fs(D-1000)}{(s-1000)} = \dfrac{5.3 \times 3100 \times (1424.63-1000)}{(3100-1000)} = \dfrac{6976670.9}{2100} = 3322.22 \, \dfrac{\text{kg}}{\text{h}}$

$= 3.32 \, \text{t/h}$

Therefore, the tonnage of dried solids pumped $= 4.402 + 2.97 + 3.32 = 10.69$ t/h

Example 10.4

In a six-hour shift, a froth flotation plant treats 2.700 tons of solids. The feed slurry contains 45% solids by weight and is delivered into a conditioning tank for treatment with reagents for 6 min before being pumped to the flotation operation. Calculate:

i. The slurry density
ii. The mass flow rate of solids in the slurry
iii. The volumetric flow rate of solids in the slurry
iv. The mass flow rate of water in the slurry
v. The volumetric flow rate of water in the slurry
vi. The volumetric flow rate of the slurry
vii. The volume of the conditioning tank required

Take the specific gravity of the solids as 2710 kg/m³.

Solution

i. The average mass flow rate of solids in the slurry $= \dfrac{2710}{6} = 450 \dfrac{\text{tons}}{\text{h}}$

$= 450,000 \, \text{kg/h}$

ii. The slurry density $= D = \dfrac{1000 \times 100s}{s(100-x)+1000x}$

$= \dfrac{1000 \times 100 \times 2710}{2710 \times (100-45)+1000 \times 45}$

$= \dfrac{149050}{194050} = 1396.55 \, \text{kg/m}^3$

iii. The volumetric flow rate of the solids in the slurry stream

$M_{vs} = \dfrac{450 \times 1000}{2710} = 166.05 \, m^3$

iv. The mass flow rate of water in the slurry stream

$$M_{vw} = M_{vs} \times W_r = 450 \times \frac{100-45}{45} = 166.05 \times 1.222 = 550\,t/h$$

Volumetric flow rate of the water in the slurry stream = mass flow rate of

water in the slurry/Specific gravity of water $= \dfrac{550 \times 1000}{1000} = 550 \text{ m}^3/h$

v. Mass flow rate of the slurry

Mass flow rate of the slurry = mass flow rate of quartz solids + mass flow rate of water = 450 + 550 = 1000 t/h

vi. Volumetric flow rate of the slurry = 550 + 166.05 = 716.05 m³/h

Alternatively, it can be determined as:

Volumetric flow rate of the slurry

$$= \frac{Mass\ flow\ rate\ of\ the\ slurry}{Density\ of\ the\ slurry} = \frac{1000}{1396.55} = 716.05 \text{ m}^3/h$$

Thus, for a treatment or retention time of 6 min, the volume of the conditioning tank required is:

$$= \frac{6}{60} \times 716.05 = 71.61\,m^3$$

Example 10.5

It is required to prepare 2 litres of slurry at 30% solids by weight. Given that the density of soilds in the slurry is 2800 kg/m³, determine the weights of the solids and water required to produce the slurry.

Solution

The slurry density $D = \dfrac{10^5\,s}{s(100-x)+1000x} = \dfrac{10^5 \times 2800}{2800(100-30)+1000 \times 30}$

$$D = \frac{28 \times 10^7}{226000} = \frac{280000}{226} = 1244.44\,kg/m^3.$$

But

$$D = \frac{mass\ of\ slurry}{volume\ of\ slurry}.$$

$$D = \frac{m_s + m_w}{v_p}.$$

where $m_s = mass\ of\ solids\ in\ the\ slurry$

$m_w = mass\ of\ water\ he$

$v_p = volume\ of\ slurry\ or\ pulp$

But

$$v_p = v_s + v_w$$

$$v_p = \frac{m_s}{2800} + \frac{m_w}{1000}$$

Therefore, we have to solve two simultaneous equations:

$$1244.44 = \frac{m_s + m_w}{0.002}$$

$$0.002 = \frac{m_s}{2800} + \frac{m_w}{1000}$$

The two equations simplify to:

$$2.49 = m_s + m_w$$

$$5.60 = m_s + 2.8m_w$$

$$m_w = \frac{3.11}{1.8} = 1.73\,\text{kg}$$

$$m_s = 2.49 - 1.73 = 0.76\,\text{kg}$$

10.5   Mass Balancing Methods

Mass balancing of plant products, assumed to be only concentrates and tailings, is required to assess the performance of a plant and control its operation for improved efficiency. It is required to know how the metal content of the ore feed is distributed in the concentrates and tailings. In the mass balancing method, only two products, concentrates and tailings, are assumed from the processing of the feed. If the flow rates of the feed, the concentrates and tailings in tons/hour are taken as F, C and T and the grades are represented as f, c and t, respectively, then mass and metal content balance are given as:

$$F = C + T \tag{10.4}$$

$$Ff = Cc + Tt \tag{10.5}$$

Solving the simultaneous equations yield:

$$\frac{F}{C} = \frac{c - t}{f - t} = ratio\,of\,concentration \tag{10.6}$$

$$\frac{c}{f} = ratio\,of\,concentration$$

The plant recovery is given by:

$$Recovery = \frac{Metal\ content\ of\ the\ concentrate}{Metal\ content\ of\ the\ feed} \times 100\%$$

Substituting for C/F from Equation 9.6, we obtain:

$$Recovery = \frac{Cc}{Ff} = \frac{C}{F} \times \frac{c}{f} \qquad Recovery = \frac{100c(f-t)}{f(c-t)}\%$$

Thus, values of recovery, ratio of concentration and enrichment ratio can be determined from the assay results only and they can be used retrospectively to assess plant performance. However, direct, real-time control can be achieved if up-to-date assay results can be continuously determined online using on-stream analysis methods.

Example 10.5

Given Table 10.1 as the performance of a shift in a mineral processing plant, complete the table using the two product formulae.

Table 10.1 should be below.

TABLE 10.1
Shift 1 Performance

Item	Weight (tons)	Assay (%)	Metal content (tons)	Metal distribution (%)
Feed (F)	260	2.8	–	–
–	–	42.5	–	–
–	–	0.17	–	–

Where the feed treated during the shift was 260 tons.

Solution

From the two product formula,

$$\frac{F}{C} = \frac{(c-t)}{(f-t)}$$

Therefore:

$$C = \frac{(f-t)}{(c-t)} \times F = \frac{(2.8-0.17)}{(42.5-0.17)} \times 260 = 16.15\ tons$$

Mass of tailings (T) = F–C = 260–16.15 = 243.85 tons
The metal content of the feed = 260 × 0.028 = 7.28 tons
The metal content of the concentrate = 16.15 × 0.425 = 6.86 tons
The metal content of the tailings = 243.85 × 0.017 = 0.42 tons

Therefore, % metal content in the concentrate $= \frac{6.86}{7.28} \times 100 = 94.23$

Therefore, % metal content of the tailing $=\dfrac{0.42}{7.28}\times100=5.77$

An observation of the three grades indicated in the Table shows that the second item should be the concentrate and the third one, the tailings

The completed table, Table 10.2, is presented below.

TABLE 10.2
The Completed Shift Table

Item	Weight (tons)	Assay (%)	Metal content (tons)	Metal distribution (%)
Feed (F)	260	2.8	7.28	100
Concentrate	16.15	42.5	6.86	94.23
Tailings	243.85	0.17	0.42	5.77

Note: The feed can also be expressed as flow rate in tons/hour.

10.6 Slurry Pump Calculations

In order to successfully pump slurries, practical methods evolved over the years by modifying the existing theories are the keys. The critical velocity of a slurry typically varies from 0.90 to 5.40 m/s as a function of the constituent solids' specific gravity, screen distribution and percentage content. For instance, tailings containing slimes are mostly pumped at velocities not less than 0.90 m/s, while gravel dredges and sands are pumped at velocities that can be up to 5.40 m/s.

The centrifugal and positive displacement pumps are the two main families of pumps. A centrifugal pump, also known as a lined pump, is used for transporting fluid exposed to highly abrasive solids. Therefore, they are produced with materials that are resistant to corrosion and abrasion. It operates by transferring the kinetic energy of the motor to the slurry using a spinning impeller. As the impeller rotates, the slurry is drawn in causing increasing velocity which moves the slurry to the discharge point. The high-pressure reciprocating displacement pumps are preferred in pumping sludge, paste and slurries of high solid contents.

In the positive displacement pumps, a fixed volume of slurry is trapped in a cavity and afterwards the trapped slurry is forced into the discharge pipe. In a centrifugal pump, the flow rate varies with pressure while it remains constant in the positive displacement pumps. Furthermore, the flow rate decreases with increasing viscosity in the centrifugal pumps due to friction losses. But the positive displacement pumps have high internal clearances that enable them to handle highly viscous slurry easily such that the flow rate increases with increasing viscosity. Efficiency of centrifugal pumps also peaks at a specific pressure, unlike the positive displacement pumps which are less affected by pressure.

Centrifugal pumps are most commonly used for the transfer of low-viscosity fluids at a high flow rate in general water supply, seawater transfer, water circulation, boiler feed, air conditioning, petrochemical, irrigation and firefighting. However, the positive displacement pumps are used to transfer high-viscosity fluids at high pressures where centrifugal pumps will fail. There are two types of positive displacement pumps-rotary and reciprocating. The rotary type operates by the rotation of the pumping elements while the reciprocating works by the back and front motion of the elements. Figure 10.1 shows a centrifugal slurry pump.[2]

A slurry flow rate is the volume of the slurry per unit of time flowing passing a point through cross-sectional area A in the pipe. Flow rate is measured in m³/s or litre/s where 1 m³/s = 1000 L/s. The flow rate, Q, through a pipe is given by:

$$Q = Av \tag{10.7}$$

where
   A = internal cross-sectional area of the pipe
   v = fluid velocity through the pipe (m/s)
   Q = fluid flow rate

Slurry piping system consists of pumps, pipelines and valves. The behavior of the slurry is determined by its solid concentrations, the hardness of the solid and the carrier fluid density and viscosity. Slurries can be homogeneous when the particles are

FIGURE 10.1   Centrifugal slurry pump.

Source: Courtesy of 911 Metallurgists (2022)

uniformly distributed in the carrier fluid such as for slurries consisting of a high content of fine particles. In heterogeneous slurries, on the other hand, the particles are not uniformly distributed. This situation is also obtained in horizontal pipes where there is a higher concentration of particles at a lower level than at the upper level. However, most industrial slurries exhibit a behavior between those of homogeneous and heterogeneous slurries, particularly when they consist of a mixture of particles of different sizes and the dominant characteristic needs to be identified for design.

In pipeline sizing, the following steps are required:

a. Determine the slurry flow rate or throughput
b. Using Equation 10.7, determine the design velocity (V)
c. Calculate the critical velocity

Clogging of the pipeline carrying a slurry will not take place if the slurry velocity is higher than some critical value. Critical velocity is the minimum velocity to keep solids in a suspended condition in a pipe. The inlet slurry velocity must be selected so that the settling of solids will not occur. The correct inlet velocity depends on the type of solids and the particle size of the solids. For instance, for fine slurries with above 200 mesh size (74 μm), the minimum inlet velocity is required to be between 1 and 1.5 m/s, while for medium-sized sand with sizes between 841 and 74 μm, the inlet velocity of 1.5–2 m/s is recommended.

The Reynolds number (Re)[3] is a dimensionless quantity and it is used to indicate the type of fluid flow pattern in a pipe is either laminar or turbulent. It is numerically equal to the ratio of inertia forces in a fluid to its viscous forces.

$$Re = \frac{Inertia\ Force}{Viscous\ Force} = \frac{\rho VD}{\mu} \tag{10.8}$$

where
   Re = the Reynolds number
   ρ = fluid density, kg/m^3
   V = fluid flow velocity, m/s
   D = the carrier pipe's internal diameter, m
   μ = the fluid viscosity, Ns/m^2

The patterns of fluid flows in a pipeline may be laminar, transition or turbulent depending on the Reynolds number value as follows:

Laminar fluid flow   up to Re = 2300
Transition   2300 < Re < 4000
Turbulent   Re > 4000

A Newtonian fluid is one which has viscous stresses due to its flow at every point to have a linear correlation to the local strain rate, that is, the rate of change of its deformation over time. Furthermore, the associated stresses are proportional to the rate of change of the fluid's velocity vector. The pipe design must also establish a pipe diameter that will transport the slurry such that no settling of solid particles will take place. If the pipe diameter is too large for a particular flow rate, solids settling will occur due to the slowing down of the slurry, while if the reverse is the case, the pump's working will be affected in its bid to counteract a large pressure drop. A deposition velocity, v_d, that is, the minimum slurry flow velocity, then has to be determined with the equation:

$$v_d = F_r \left[2gD_i \left(\frac{\rho_s - \rho_w}{\rho_w} \right) \right] \tag{10.9}$$

where
v_d is the settling velocity in m/s
F_r is the Froude's number for the slurry
g is the acceleration due to gravity
D_i is the pipe's internal diameter
ρ_s is the density of the solids in the slurry
ρ_w is the density of water

Non-Newtonian slurries may have higher deposition velocities than Newtonian slurries.[4-7]

Example 10.6

Consider a homogeneous slurry with a flow rate of 1.5 litre/s, pipe with an internal diameter (ID) of 240 mm, a viscosity of 65 cp and a pulp density of 2700 kg/m³.
Determine:

a. The Reynolds number for the fluid flow
b. the critical velocity (Vc) of the slurry if it is Newtonian

Solution

The volumetric flow rate
$F = Av$
But 100 cm = 1 m and therefore, $10^6 \, cm^3 = m^3$
Therefore, $1 cm^3 = 10^{-6} \, m^3$ and 1 litre = 1000 cm³. Therefore,
Slurry flow rate = 1,500 cm³ $=1500 \times 10^{-6} \, m^3 = 0.0015 m^3$
$F = 0.0015 \times 10 = 0.015$ m³/s

$$A = \frac{\pi D^2}{4} = \frac{22 \times 0.24^2}{7 \times 4} = \frac{1.2672}{28} = 0.0453 \, m^2$$

Therefore

$$V = slurry \, velocity = \frac{F}{A} = \frac{0.015}{0.0453} = 0.331 \, m/s$$

Note that 1 cp = 0.01 poise

$$R_e = \frac{2700 \times 0.33 \times 0.24}{0.65} = 328.98$$

But $R_e = \dfrac{2700 \times 0.33 \times 0.24}{0.65} = 328.98$

The fluid flow is thus turbulent

For a Newtonian fluid, the critical Reynolds no. is 2100

$$Critical\,Velocity = \frac{Reynolds\,No. \times Viscosity}{Solids\,density \times pipe\,ID} = \frac{2100 \times 0.065}{2700 \times 0.240} = \frac{0.050556}{0.240}$$

$Critical\,Velocity = 0.21$ m/s

The velocity to be selected must be such that (V – Vc) > 0.3 m/s, that is, V>0.51 m/s

Example 10.7

A homogeneous slurry with a volumetric flow rate of 85 litre/s is to be carried in a pipe 650 m long and 190 mm internal diameter (ID), to determine the design velocity required.

Solution

The volumetric flow rate, $F = \dfrac{85L}{1s} = \dfrac{85 \times 1000\,cm^3}{1s}$

But 100 cm = 1 m and therefore, $10^6\,cm^3 = m^3$

Therefore, $1\,cm^3 = 10^{-6}\,m^3$ and

$$F = \frac{85 \times 1000 \times 10^{-6}}{1s} = 0.085 m^3/s$$

The internal cross-sectional

$$A = \frac{\pi D^2}{4} = \frac{22 \times 0.19^2}{7 \times 4} = \frac{0.7942}{28} = 0.028\,m^2$$

$$V = \frac{F}{A} = \frac{0.085}{0.028} = 3.04$$

Thus the average pipeline velocity is 3.04 m/s.

References

1. Wills, B.A., and Napier-Munn, T.J. (2006): *Mineral Processing Technology*, 7th Edition. Amsterdam: Elsevier Science and Technology Books.
2. 911 Metallurgists (2022): *Centrifugal Slurry Pump* https://z4y6y3m2.rocketcdn.me/equipment/wp-content/uploads/2018/01/centrifugal-slurry-pump-10.jpg, Accessed 27th April, 2022).
3. BYJU'S (2022): *Reynolds Number—Definitions, Formulas and Examples—BYJU'S* (www.byjus.com, Accessed 11th March, 2022).
4. Satish Lele Slurry Piping (SLSP) (www.svlele.com/piping/slurry_piping.htm, Accessed 16th June, 2020).

5. The Engg ToolBox (www.engineeringtoolbox.com/slurry-transport-velocity-d_236.html, Accessed 3rd September, 2020).
6. QDTHY Casting (www.qdthyhcasting.com/product_centrifugal-pump-oem-component-spare-parts_26033.html, Accessed 3rd September, 2020).
7. Yongyi Pumps (https://yongyipump.en.made-in-china.com/product/IvQnOyulAqYN/China-Positive-Displacement-Anti-Corrosive-Design-Centrifugal-Slurry-Pump-Easy-Maintenance.html, Accessed 3rd September, 2020).

11 Mineral Processing Methods

11.1 Introduction

There are four categories of mineral processing methods to concentrate ore minerals based on their physical and physico-chemical properties. The methods are designed to exploit the differences in the physical, physico-chemical and surface properties of mineral particles such as density, magnetic susceptibility, charge conductivity and surface tension characteristic of mineral particles in water. The techniques that are commonly used are:

1. Gravity separation or concentration technique
   The method exploits the difference in the density of mineral particles in the ground ore and other factors such as size and shape to separate them.
2. Magnetic separation technique
   This method uses the difference in the magnetic susceptibility of mineral particles in the ore to separate them.
3. Electrostatic technique
   This technique uses the difference in electrical chargeability or the charge-retaining ability of different mineral particles to separate them.
4. Froth flotation
   The method exploits the difference in the phyisco-chemical surface properties of different mineral particles to separate them.

The enrichment process in mineral processing involves washing and separation. Washing is usually applied to coal, industrial minerals, sand and gravel with sizes ≥ 1 mm. Washing is carried out using log washers, aquamator separators, wet screens, tumbling scrubbers and attrition scrubbers. Separation is used to enrich and hence to relatively decrease the associated gangue minerals in metallic ores and high-grade industrial minerals that have been ground ≤ 1 mm to liberate the mineral values they contain. Separation is carried out by gravity, magnetic and electrostatic separation as well as by froth flotation described earlier.

In wet screen washing, water is sprayed on the material carried on a screen at a low or high volumetric flow rate, low (3–6 bar) or high pressure (70 bar) to remove dirts and slimes. The material may be an aggregate, raw ore, sand or gravel. For materials laden with high content of sticky clay, a washing drum called a tumbling scrubber is used. It rotates at a medium speed to cause the scrubbing of solids against solids for clay removal. The capacity ranges from 8 to 120 tph.[1, 2]

DOI: 10.1201/9781003323433-11

11.2 Gravity Concentration Methods

11.2.1 Introduction

They are preferably used for coarse particles but can also be used for fine particles down to those with sizes less than about 50 μm. The areas of applications in the mineral industry include:

1. They are preferably used for heavy sulphide minerals such as galena with a specific gravity of 7.5, but can also be used for particles of light minerals such as coal with a low specific gravity of 1.3 in comparison to galena.
2. It is the choice method to process iron, tungsten and tin ores. Expensive methods such as froth flotation are preferably and more profitably used to treat gravity concentrates than to treat the original ores. Therefore gravity separation is a valuable pre-concentration method.

11.2.2 Advantages of Gravity Methods

Advantages of gravity methods include:

1. It is relatively cheap because the benefit of gravity can be exploited in the downward flow of the slurry
2. The process is also simple in comparison with some other methods
3. They are more ecologically and environmentally acceptable than methods such as froth flotation that uses chemical reagents
4. It has been found that most mineral values in a complex ore can first be gravity pre-concentrated before completing the concentration with more expensive methods
5. The possibility of using modern gravity techniques to treat fine-sized ores as fine as 50 μm

11.2.3 Principles of Gravity Separation

The gravity methods are designed to cause the ore particles to fall under gravity in a carrier fluid either by slanting in shaking tabling or vertical descent in a spiral conduit in a spiral separator. As the particles fall under gravity, they encounter a viscous resistance of the carrier fluid to their motion, which depends on each particle's density, size, shape as well as surface tension and thus the particles experience differential movements. The effect of gravity falling on the particles is accentuated by actions such as shaking in shaking tabling or spiral motion in a spiral separator. The particles' differential movements cause them to separate out at different points at the receiving end-point of the system.

11.3   Factors that Affect Gravity Concentration's Efficiency

11.3.1   Density of Ore Minerals

For efficient separation between mineral value particle and gangue mineral constituents, a pronounced or clear difference in density between the two is required. The concentration criterion that expresses the degree of efficiency based on density difference is given by

$$Cc = \frac{D_h - D_f}{D_l - D_f} \qquad (11.1)$$

where
   D_h is the heavy mineral density, assumed to be the mineral value
   D_l is the light mineral density, assumed to be the gauge mineral
   D_f is the carrier fluid density
   If Cc > 2.5, either positive or negative, then the gravity-based separation process is considered relatively efficient or easy
   As the value of Cc decreases below 2.5, the efficiency of the gravity separation also decreases

11.3.2   Ore Particle Size

In general, separation efficiency increases as the particle size of the ore to be treated increases. If the particle is sufficiently coarse such that the terminal velocity can be predicted with Newton's law, then ore particle density is the predominating factor for its gravity separation and Equation 11.1 can be used to predict the ease and efficiency of the separation. For particles that are too fine such that Newton's law does not apply, surface friction factor predominates and gravity separation will not be efficient even when the density advantage exists. The effects of particle size on the efficiency of gravity and other mineral processing methods are shown in Figure 1.8 reported by Wills and Napier-Munn (2006)[1]. Table 11.1 shows the effects of the concentration criterion on the feasibility of gravity separation.[3]

11.3.3   Slimes

Slimes are ultra-fine particles less than about 10 µm in size. They increase the viscosity of the slurry and reduce the sharpness of gravity separation. Ground ores with slimes should be de-slimed using hydrocyclone.

TABLE 11.1

Effects of Concentration Criterion on the Feasibility of Gravity Separation

Concentration Criterion	Suitability for Gravity Separation
Cc >2.5	Separation of particles as very fine as 75 µm size is considered easy
1.75 < Cc < 2.5	Separation of particles as coarse as 150 µm and not below is considered possible
1.5 < Cc < 1.75	Separation of particles as coarse as 1700 µm and not below is considered possible
1.25 < Cc < 1.5	Separation of particles as coarse as 6350 µm and not below is considered possible
Cc < 1.25	Separation at any size, no matter how coarsely liberated, is considered impossible

11.3.4   Correct Water Ratio

The water ratio determines the pulp density and all gravity concentration methods have optimum pulp density specifications for efficient separation. A little deviation from this optimum density will adversely affect the gravity process.

11.3.5   Presence of Sulphide Minerals

The presence of sulphide minerals adversely affects the operation of spirals and tables. If sulphide is present, it can be removed by flotation before gravity separation. If the particle size of the sulphide-bearing ore to be processed is less than 300 µm, it can be subjected to flotation first. Afterwards, the concentrates can be treated with spiral and tables to separate other values.

11.3.6   The Presence of Useful Valuable Contaminants

There are usually some valuable contaminants in ore minerals that the gravity concentration method cannot separate. Hence, gravity concentration cannot be considered a final process. For example, cassiterite ore usually contains valuable contaminants such as magnetite and wolframite after gravity separation. These two valuable contaminants can be removed using a two-pole magnetic separator that will remove the ferromagnetic magnetite at the first weak pole and the paramagnetic wolframite at the second strong pole. The diamagnetic cassiterite will then pass through the poles without being selected by the poles and is collected as the final concentrate with low concentrations of the two contaminants.[1]

11.4   Gravity Concentrators

There are several types of separators designed to separate mineral value(s) from gangue minerals based on their differential movements when falling under gravity. These include jigs, cones, wet tables, spirals, wet sluices, tilting frames and mozley tables. The efficiency of the techniques depends mainly on the size of the ore feed as shown by Wills and Napier-Munn (2006) in Figure 1.8. Figure 1.8 shows that jigs will only begin to work efficiently with ore feeds of about 150 μm size and the efficiency will increase as the feed size increase to about 1000 μm. The spirals will only be efficient from an ore feed size of 80 μm and becomes increasingly efficient as the feed size increases to about 100 μm from where the efficiency remains almost constant until about 500 μm feed size and then the efficiency begins to decrease to about 1000 μm size. On the other hand, the tilting frame is efficient from an ultra-fine size of about 10 μm and the efficiency increases up to about 90 μm feed size and begins to decrease until it becomes zero at about 100 μm.

11.4.1   Jigging: Gravity Separation

In jigging different materials of varying density in an ore are sorted in a fluid by making them to be stratified. The sorting occurs due to a bed of particles moving as the bed gets an inflow of fluids on an intermittent basis by the fluid pulsating vertically in a plane. The stratification makes the charge particles get settled in layers with decreasing density from the bottom to the top.

Jigs are used in the gold, barytes, tungsten, coal, cassiterite and iron ore industries. In jigging, the separation of minerals of different densities is achieved on an ore bed made fluid by a pulsating current of water and hence causing the ore particles to stratify. In typical laboratory practice, about 100 g of the ore with known particle size consist is stored in a tray to form the feed or head material for the jigging operation. Steel balls are spread to form a thin bed layer on the screen of the mineral jig as a bedding material. The spigot of the hutch compartment is plugged with rubber cork and water is added to cover the ragging in the feeding compartment. The head or feed is delivered into the feeding compartment. Feed material mixed with water at average dilutions (1:1) is added to the jig and the jigging process with medium stroke and speed is allowed for 4 min. On the completion of each jigging processing, the product is collected as the underflow by opening the spigot of the hutch compartment. The two products, underflow and overflow, are dried, weighed and recorded.[3]

11.4.2   The Jigging Process

Jig induces the separation of particles based on differential acceleration, hindered settling and consolidated trickling using the pulsating of fluid at some frequency and amplitude. The separation involves the following steps:

a. Particles to be separated are in a mixed pile
b. As the water level rises, the bed layer is lifted
c. The sedimentation and stratification of the particles in water occur
d. When the water level reduces, the heavy mineral particles settle at the bottom

Figure 11.1 shows a jigging concentrator.[4]

FIGURE 11.1   Gravity jig.

Source: **Courtesy of 911 Metallurgist (2022a)**

11.4.3   Shaking Tabling

When water flows over an inclined surface, the water closest to the surface becomes retarded by the friction it encounters from the water absorbed into the surface and thus the velocity of the water increases towards the surface. As a result of this, if ore particles are introduced into the water, the larger particles have higher velocity than the smaller ones that are submerged in the slower-moving region near surface water. Furthermore, denser particles will move at lower velocities than lighter ones and lateral displacements of the particles will occur and cause them to move in different directions to the collecting endpoints.

In typical laboratory practice, a slurry with 25% solids by weight was produced was introduced at the feed box and distributed in the wash water along the feed side from the launder. The table was then vibrated longitudinally by the shaking mechanism, using a slow forward stroke and a rapid return which caused the mineral particles to 'crawl' along the deck parallel to the direction of motion. The minerals were thus subjected to two forces, forces due to the table motion and that at right angles to it, due to the flowing film of water. Consequently, the particles move diagonally across the deck from the feed end and fan out on the table, the smaller, denser particles riding highest towards the concentrate launder at the far end. While the larger lighter particles are washed into the tailings launder, which runs along the length of the table. An adjustable splitter at the concentrate was used to separate the product into three fractions which are a high-grade concentrate and two middling fractions.

The concentrate was further shaken continuously for another nine times, weighing the products as wet after every shaking. The final products were dried separately and then weighed.[3]

Akande and Adeleke (2019)[5] reported the shaking tabling processing of the sand residue from Ondo tar sand. The sand slurry with 25% solids by weight was introduced at the feed box of shaking table model no. ED808148566Q. The table was then vibrated longitudinally by the mechanism, using a slow forward stroke and a rapid return which caused the mineral particles to 'crawl' along the deck parallel to the direction of motion. The particles moved diagonally across the deck from the feed end spread out on the table, the smaller, denser particles moving towards the concentrate launder, while the larger lighter particles were washed into the tailings launder. An adjustable splitter at the concentrate end was used to split the product into three fractions—a high-grade concentrate and two middling fractions. The concentrate fraction was further shaken another nine times. The products were weighed wet after each shaking. The final products were dried separately and then weighed.

11.4.4 Humphrey Spiral

Spirals are commonly used in processing heavy mineral sands and chrome-bearing sands, that is, unconsolidated minerals with specific gravity (sg) of between 4 and 5.5. Heavy mineral sands include monazite (sg 4.6–5.7), rutile (sg 4.23), ilmenite (sg 4.79), leucoxene (altered ilmenite, sg 4.3) and limonite (sg 2.9–4.3), zirconia (sg 5.68) and chromite (sg 4.5–5.09). The market value of trading in these commodities currently amounts to billions of dollars. To underscore the importance of these materials, China's consumption of zirconia increased by an annual average of 17.20% between 1990 and 2008.

The spiral has a helical conduit that is semi-circular in cross-section. It is designed to receive ore pulp with sizes as coarse as 3 mm but not smaller than 75 μm in size at a pulp density ranging between 15% and 45% by weight. The ore pulp is introduced at the top and it flows downwards spirally through the helical conduit. The combined effects of the centrifugal force, ore particles' differential settling rates and interstitial trickling of the particles through the particle bed cause the mineral particles to stratify so that the denser particles get separated in a band along the inner edge of the stream as the concentrate. Figure 11.2 shows the Humphrey spiral and Figure 11.3 presents a modern spiral concentrator.[6]

When spiral treats a slurry, the lighter gangue minerals get more readily suspended than the heavy value minerals and develop a higher tangential velocity and hence are carried to the outer rim of the trough. The wash water which is added along the inside rim aids in washing off the lighter gangues. The formation of bands of the heavy mineral occurs in the inner rim after some turns in the spiral. There are spaced ports along the inner rim to collect the valuable mineral, while the gangue is taken at the bottom. It is usually operated industrially in a five-stage circuit consisting of a slime removal stage and rougher, scavenger, cleaner and re-cleaner stages. The ore feed is firstly de-slimed in a cyclone and then taken to a sump for water addition in addition to decreasing the % solids to 35. Figure 11.4 shows a labelled spiral concentrator.[6]

FIGURE 11.2   The original Humphrey spiral.

Source: **Courtesy of 911 Metallurgist (2022b)**

The modern TOMINE spirals are made of fibreglass lined with wear-resistant resin and emery. The equipment can be used to effectively enrich particles of coal, iron ore, cassiterite, chromite, ilmenite, monazite, rutile and ores of tungsten, zinc, gold, and zircon of sizes 0.3–0.02 mm. They are also used for other low-grade ores or non-metallic minerals. The slurry is gravity fed into the spiral. The spiral concentrator's separation by density is based on the primary downward flow and circulating secondary flow patterns developed due to the slowing down of the slurry near the trough surface as the slurry charge descends along the helical trough surface that spiral around a central axis downward to the discharge end.[7]

The primary flow causes the particles to travel downwards towards the discharge point, while the secondary flow causes the density separation by carrying the lighter particles to the outer trough, which is like helical sluices. The heavy particles are

FIGURE 11.3   Modern spiral concentrator.

Source: Courtesy of 911 Metallurgist (2022b)

moved to the slow-moving inner portion of the slurry stream from where they are taken away through concentrate or middling discharge outlets. The spiral's adjustable splitters allow any portion of the concentrate or middling to be diverted through the outlets. The splitter can be adjusted from the central column towards the end of the trough. The tailings are discharged through the lower end of the spiral. The main spiral operating parameters are % solids at 15 to 45% and splitter setting. Spirals are commonly used to separate fine coals from gangue mineral matter. Figure 11.5 shows the spiral discharge splitters.[6]

In coal spiral concentration, the less dense, coal particles will stay suspended in water as the coal slurry spiral downwards. The heavy discard particles will sink to the bottom of the trough where the drag due to the frictional force slows down the particles and moves them towards the central column. When the coal slurry gets to the bottom of the spiral, the coal particles are separated into the product towards the outside of the trough, while the heavy gangue particles are concentrated towards the central column.

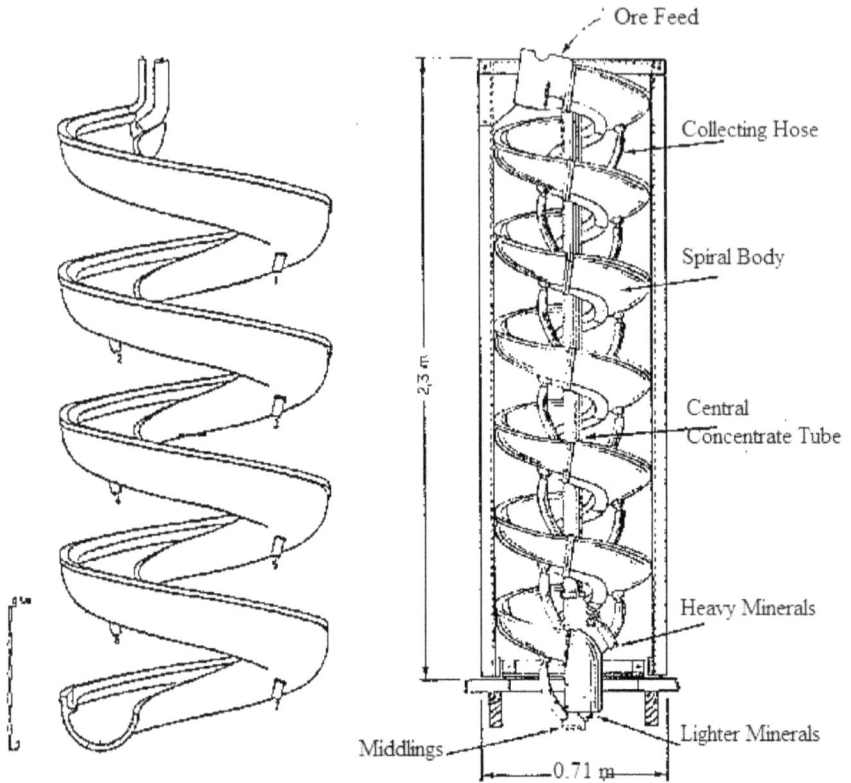

FIGURE 11.4   A labelled spiral concentrator.

Source: **Courtesy of 911 Metallurgist (2022b)**

The adjustable splitter at the bottom of the trough is manually set to separate coal from the gangue discards. Spirals are also used in the chrome industry to treat fine chrome sands <1 mm in size. The feeds are first processed in cyclones and hydrosizers before delivery to spirals.[8, 9] The advantages of spiral concentration include:

a. Low capital cost
b. It has no moving part that requires energy input and thus the operating cost is low
c. The maintenance requirements are low
d. It is easy to operate
e. It is not too sensitive to changes in slurry density

The disadvantages, however, include:

a. It cannot treat ultra-fine ores such as those containing <100 μm particles
b. The quality of the product obtained cannot be readily changed
c. It is drastically affected by the feed quality

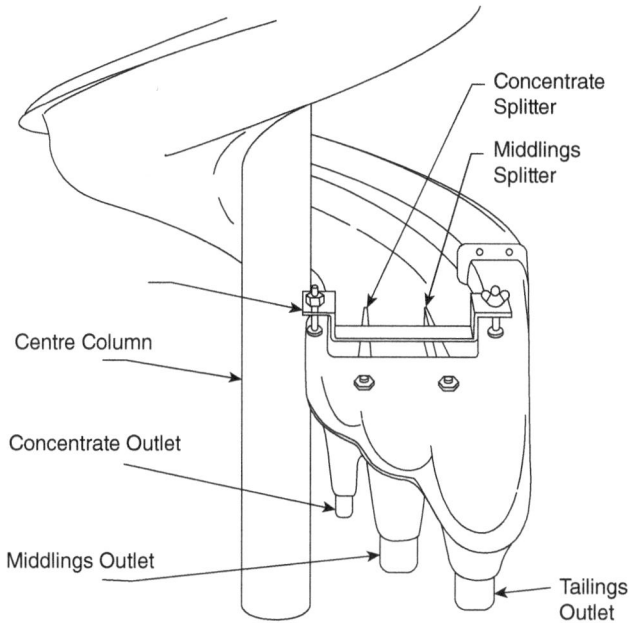

FIGURE 11.5   The spiral discharge splitters.[6]

Source: **Courtesy of 911 Metallurgist (2022b), adapted**

Figure 11.6 shows the diagram of the centre column of the spiral.[6]

The product concentrate can be collected into a container, allowed to settle and the water decanted to obtain a wet product for drying. The spiral is also used in the mill grinding circuit to remove and concentrate coarse gold particles from the re-circulating load. For the efficient operation of a spiral, the input parameters of percentage solids, the flow rate and ore feed size consist must be constant. The pulp should not be too high in order to allow the particles to be sufficiently mobile in the concentrate and middling streams. When the spiral is fed from the hydrocyclone underflow discharge box, its two splitter arms can be adjusted to vary the feed volume into the spiral.

11.4.5   Falcon Gravity Concentrator

The Falcon Concentrator is used for gravity separation at fine-size fractions. It is an enhanced gravity concentrator and it has a bowl that spins at a fast rate. The bowl takes in the slurry feed from the bottom and it drains the slurry in a thin flowing film at its wall by applying a centrifugal force. The slurry is taken directly into the rotating bowl, while a variable speed drive regulates the gravitational force equivalent, G-force where 1 G force refers to a force that produces an acceleration of 9.8 m/s. The Falcon SB concentrator's run time may be 5 min or more and the concentrator uses a dynamic braking system to quickly slow down the bowl, rinse out the concentrate

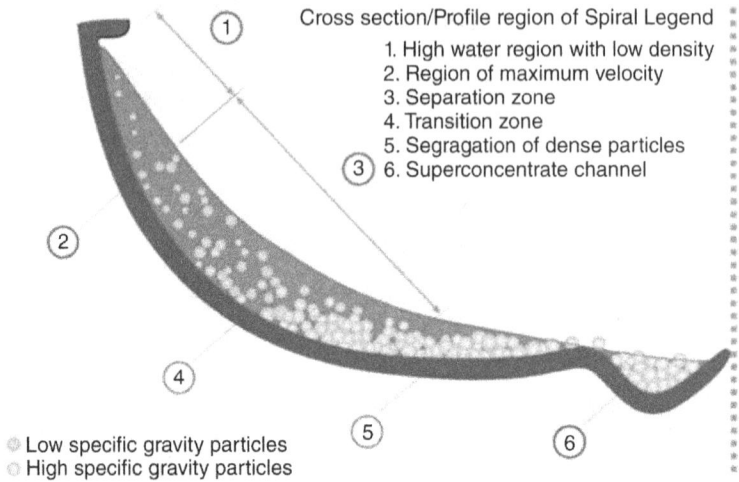

Cross section/Profile region of Spiral Legend
1. High water region with low density
2. Region of maximum velocity
3. Separation zone
4. Transition zone
5. Segragation of dense particles
6. Superconcentrate channel

Low specific gravity particles
High specific gravity particles

FIGURE 11.6   A diagram of the centre column of the spiral.

Source: **Courtesy of 911 Metallurgist (2022b), adapted**

and then return to its full operational speed. It is equipped with a variable drive system and it can thus operate from 5 to 200 G force. Tailings are received at the bottom outlet pipe of the unit while the concentrates are taken at the concentration ring at the end of each test. Areas of application include the concentration of ores of gold, platinum group metals, mineral sands, chromite, tin, tantalum, tungsten, iron and cobalt. Figure 11.7 shows a Falcon concentrator.[10]

11.5   Knelson Gravity Concentrator

The Knelson gravity concentrator (KGC) is a vertical axis bowl-type continuous discharge centrifugal concentrator that uses a fluidized bed to concentrate ores. It is now commonly used in any gold processing plant to recover free gold particles from the grinding circuit and to obtain gold content of ores amenable to gravity recovery. It uses an enhanced gravitational force coupled with a fluidization process to recover even ultra-fine-sized particles at a micro-scale. It has a series of fluidization holes through which water is first introduced into the rotating concentrating cone. The feed slurry is introduced into the cones using a stationary feed tube. By the time each cone is filled with slurry, a concentrating bed is created and particles of high density are retained in the cone and then conducted into the concentrate launders by a flushing action. The duration of the concentrating process can be less than one minute.

The Knelson gravity concentrator (KGC) is also used for gravity separation at fine-size fractions The KGC operation is based on the fluidization principle. The concentrator receives water to fill it through some fluidization holes. As the water-filled concentrator receives slurry into it, particles with high-density attempts to pass through the fluidized bed, and finally get trapped in the middle of the individual

FIGURE 11.7   Falcon concentrator.

Source: **Courtesy of 911 Metallurgist (2022c)**

rings of the cone. However, particles with lower density cannot pass through the fluidized bed and are then displaced over the top of each ring to create a stream of continuous tails that overflow the inner bowl. The concentrate is received in a fluidized
concentrate ring, whereas the tailings with lower density are washed into the cone
discharge rim. Figure 11.8 shows a Knelson Concentrator with its component parts.[11]

The Knelson concentrator has two rings located at the top of the bowl through
which fluidized water is forced into the concentrator. High-density particles get
settled against the fluidized water and are taken away by valves placed along the
circumference. When around 60 G's of force is applied, a capacity of up to 100 tph
is obtained.

11.6   Industrial Spiral Concentration

Mankge et al. (2015)[12] reported the spiral processing of a chromite slurry that assayed
20% Cr_2O_3 with 15–45% solids in the size range of 25 μm to 3 mm. The slurry was
introduced at a flow rate of about 430 m³/h at the top of the spiral. As the slurry
flowed spirally downwards, the particles became stratified due to the combined effect
of centrifugal force, the differential settling rates of the particles and the effect of
interstitial trickling through the flowing particle bed. The wash water added at the
inner edge of the stream flowed outwardly across the concentrate band. The width

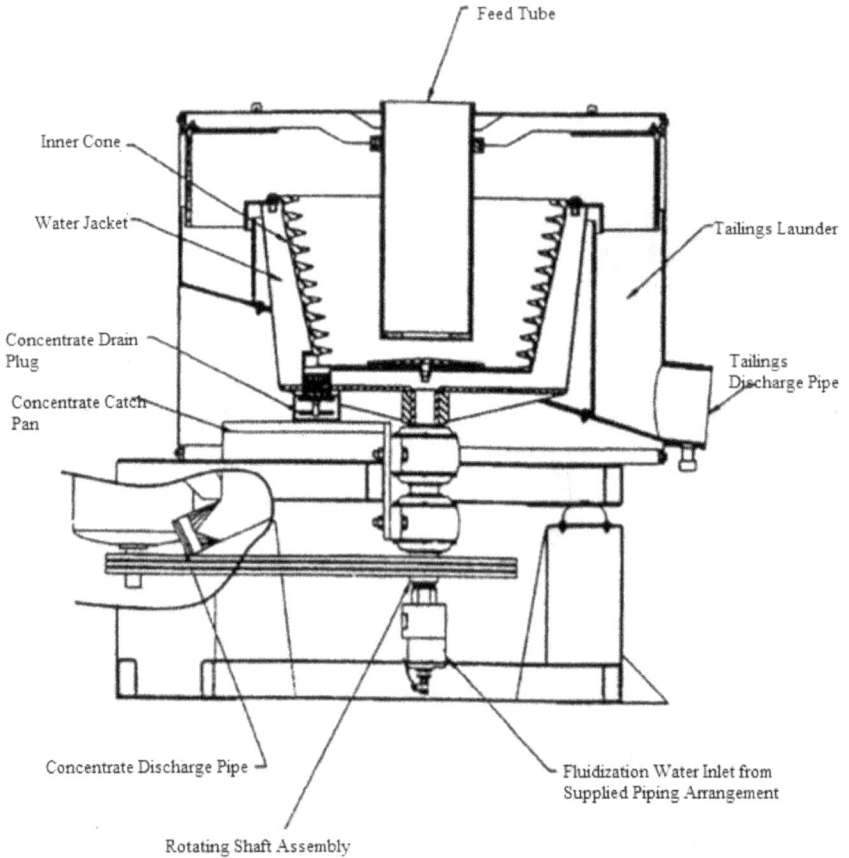

FIGURE 11.8   Knelson Concentrator with its components.

Source: Courtesy of 911 Metallurgist (2022d)

of the concentrate band removed at the ports was controlled by adjustable splitters which separated the concentrate, middling and tailings.

References

1. Wills, B.A., and Napier-Munn, T.J. (2006): *Mineral Processing Technology*, 7th Edition. Amsterdam: Elsevier Science and Technology Books.
2. Metso (2022): *Basics in Mineral Processing* (https://vdocument.in/basics-in-minerals-processingmetso.html, Accessed 23rd June, 2022).
3. University of Alaska, Fairbanks (UAF) (2020): *Mining Mill Operator Training: Lesson 2: Classifying Cyclones* (https://millops.community.uaf.edu/amit-145/, Accessed 12th August, 2020).
4. 911 Metallurgist (2022a): *Jigging Machine* (https://z4y6y3m2.rocketcdn.me/wp-content/uploads/2017/02/jigging_machine.png, Accessed 22nd May, 2022).

5. Akande, R., and Adeleke, A.A. (2019): Multistage Gravity Beneficiation of Rutile in a Tar-Free Sand Residue. *Hungarian Journal of Industry and Chemistry*, 47(2), 25–30. (https://doi.org/10.33927/hjic-2019-17).
6. 911 Metallurgist (2022b): *Gravity Spiral Concentrator Working Principle* (www.911 metallurgist.com/blog/gravity-spiral-separator-working-principle, Accessed 15th May, 2022).
7. TOMINE (www.topmie.com/spiral-concentrator/, Accessed 5th June, 2020).
8. Anon (2013a): *Students' Experiential Training on Spirals*. Vanderbijlpark, South Africa: Vaal University of Technology.
9. Anon (2013b): *Students' Experiential Training at SAMANCOR Easter Chrome Mine*. Vanderbijlpark, South Africa: Vaal University of Technology.
10. 911 Metallurgist (2022c): *Falcon Concentrator* (www.911metallurgist.com/equipment/ laboratory-falcon-concentrator, Accessed 6th July, 2020).
11. 911 Metallurgist (2022d): *Knelson Concentrator* (https: www./911metallurgist. com/knelson-concentrator/0, Accessed 5th May, 2020).
12. Mankge, T.M.C., Adeleke, A.A., and Mendonidis, P. (2015): Optimization of Chromite Recovery from Cyclone-Spiral Circuit Concentration of a UG2 Ore. *Journal of Engineering Research*, 20(1), March.

12 Dense Medium Separation

12.1 Introduction

Dense Medium Separation (DMS), also known as the Sink and Float process or Heavy Medium Separation (HMS) refers to the separation of minerals of different specific gravities (sg) using a dense liquid or a heavy medium.[1] Uses of heavy medium separation include:

  a. The pre-concentration of minerals
     To pre-concentrate ore minerals by removing coarse gangue minerals before further grinding for final concentration to recover mineral values.
  b. In coal preparation, separate the light coal particles with a density of 1.3 from shale with a density between 1.77 and 3.33 or from denser coals laden with high ash. Shale is a mud consisting of clay minerals in flake forms and very small broken pieces of minerals, particularly quartz and calcite.

12.2 The Principles of Dense Medium Separation

Heavy liquids of specific densities are prepared to separate heavy minerals from light minerals. The mineral particles are mixed with the liquid of known density. The mineral particles that are less dense than the liquid will float on it as a float product, while those with higher density will sink into the bottom. Figure 12.1 shows the principles of dense medium separation.[2]

The ease or otherwise of the dense medium separation can be predicted by the formula:

$$D_d = \frac{\left(D_{hm} - D_{hl}\right)}{\left(D_{lm} - D_{hl}\right)}$$

where
D_{hm} = density of the heavy mineral
D_{hl} = density of the heavy liquid for the dense medium separation
D_{lm} = density of the main light mineral gangue in the ore

For the Dd values of +2.50, 1.75–2.50, 1.50–1.75, 1.25–1.50 and <1.25 the dense medium separation is easy, possible, difficult, very difficult and not possible, respectively.

Feed of Sized Particles

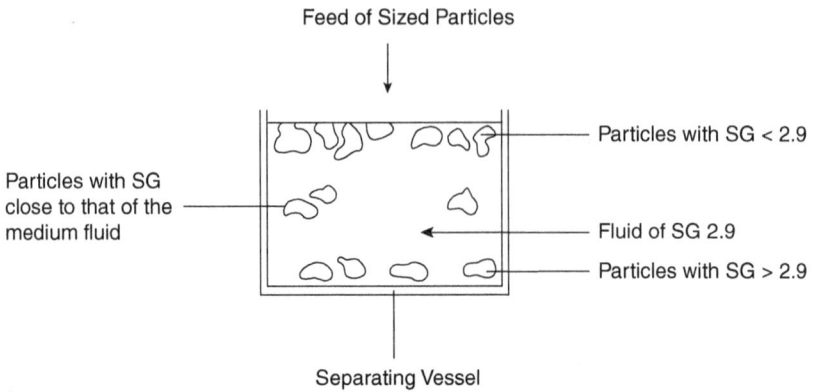

FIGURE 12.1   The principles of dense medium separation.

Source: **911 Metallurgist (2022a), adapted**

12.3   Types of Heavy Liquids

There are three types of heavy liquids as follows.

12.3.1   Dense Organic Liquid Chemicals

Dense liquid organic chemicals are dense because of their long chains. Examples are tetrabromoethane (TBE), carbon tetrachloride and bromoform with specific gravities of 2.20, 1.58 and 2.89, respectively. Organic liquids are generally toxic and volatile. Table 12.1 shows the heavy liquids to be used for dense medium separation within certain specific gravity ranges.[2]

12.3.2   Dense Inorganic Aqueous Solution

Dense inorganic aqueous solutions such as solutions of sodium polytungstate and Clerici in water. Sodium polytungstate $3[(Na_2WO_4).9WO_3.H_2O)]$, is a non-toxic, non-flammable water-based heavy liquid with a specific gravity of 5.47 and a solubility greater than 1000 g/litre at 20°C. Clerici solution is a solution of equal parts of thallium formate and thallium malonate. It is the heaviest water-based liquid. Clerici solution is free-flowing and has a high density of 4.25 g/cm^3 at 20°C and 4.36 g/cm^3 at 25°C. Unlike organic liquids, it is non-toxic, virtually non-volatile and has a low viscosity.

12.3.3   Suspensions of Ore Particles in Water

Suspensions of heavy mineral particles in water are also used as heavy media. Suspensions are produced by dissolving finely ground dense solid particles in water to form about 15% by weight of the solution. The particles of ferrosilicon (sg 6.7) and magnetite (sg 5.15) are commonly used. Milled ferrosilicon (with Fe ≥ 82% and

TABLE 12.1
Heavy Liquids for Laboratory Dense Medium Test

S/N	Specific Gravity Range	Liquid used (Specific Gravity)
1	0.80–1.60	Benzene (<1)
2	1.60–2.89	Bromoform (2.89)
3	2.40–2.96	Tetrabromoetane (2.96)
4	2.96–3.31	Methylene iodide (3.31)
5	3.31–4.00	Clerici's solution (4.00)

15–16% Si) and ground 30–95% < 45 µm sieve aperture. The mixture must be kept agitated without shear so that the particles will remain suspended in the medium. Such liquids with about 15% particles by volume are called Newtonian's fluids because they obey Newton's law. A Newtonian fluid is a fluid that obeys Newton's law of viscosity, that is, the viscosity is independent of the shear rate. In a Newtonian fluid, the viscous stresses arising from its flow, at every point, are linearly correlated to the local strain rate.

The viscosity of non-Newtonian fluids increases when shear stress is applied. Examples of non-Newtonian fluids are suspensions with solid particles greater than 15% by weight of the solution, gels and colloids, while water, mineral oils, gasoline, organic solvents and honey are examples of Newtonian fluids. Dense Medium Separation technique offers the following advantages over other gravity processes:

a. The ability to efficiently make sharp separations between minerals in an ore at any given density even when the ore contains a high proportion of near-density minerals
b. The selected separating density can be easily altered as desired to meet varying specifications

The conditions that favor the use of the Dense Medium Separation technique are:

a. When an ore contains components such that after a comminution process to achieve a suitable level of liberation between the components, the specific gravities of their particles become sufficiently different to be able separate those which can repay the extra cost of further processing from those that cannot.
b. It also applies to an ore that has its mineral value component(s) in coarse aggregates with sizes that are larger than 500 µm.
c. The process is most commonly used when after the comminution process, the resulting mineral particles in an ore are coarse and have a density difference between them. This is because the efficiency of separation between particles of different densities decreases with decreasing particle size due to the fact that fine particles exhibit slower rates of settling than coarse ones.

d. When the ore particles to be treated are preferably larger than about 4 mm in diameter in size, the efficiency of separation is good to an extent that separation can even be effective for particles with a difference in specific gravity of 0.1 or less.

12.4   Coal Sink and Float Test

Sink and float tests are carried out on run-of-mine coals to determine their washability characteristics. Washability is the process of removing mineral matter from run-of-mine coals so that the end product meets the customer's requirement. The coal is crushed and screened into size fractions and sink and float tests are carried out on the different size fractions. Zinc chloride solution can be used as the medium since it can give a relative density intermediate between that of the coal and the mineral matter content. Relative densities between 1.30 and 1.80 in the ranges 1.30–1.35, 1.35–1.40, . . ., 1.75–1.80 are used and mass recovered, and yield percent, moisture, calorific value, volatile matter, ash and sulphur in weight % are determined for coal floats for each range.

The samples are first tested on the lowest specific gravity of 1.30. Coal being less dense than the mineral matter will float, while the mineral matter will sink. Washability curve shows the relationship between ash content and the amount of sink or float produced at any relative density and enables a choice of the relative density to beneficiate raw coal.[3]

12.5   Dense Medium Industrial Separation

In the DMS industrial separation of coal from its ash, magnetite particles suspended in water are used as the separating medium. The density of the medium is adjusted to be between that of the coal and its ash component and corresponds to a density to obtain a desired coal concentrate grade in terms of ash content. The light coal float is taken away by a flight conveyor while the heavy sink ash materials are moved away by a chain conveyor. Figure 12.2 shows a dense medium cyclone.[4,5]

FIGURE 12.2   Dense medium cyclone.

Source: Courtesy of 911 Metallurgist (2022b)

12.6 Liquid Mixing Equation

Consider two liquids A and B with a specific gravity of ρ_a and ρ_b respectively. If a total volume of V is to be prepared at a density ρ from two liquids, determine the volumes of A and B to be combined.

Solution

Let the volume of liquid A = V_a
Let the volume of liquid B = V_b

$$V_a + V_b = V \tag{12.1}$$

$$\frac{\rho_a V_a + \rho_b V_b}{V} = \rho \tag{12.2}$$

Solving the simultaneous equations will yield Va and Vb

$$V_b = \frac{V(\rho - \rho_a)}{\rho_a - \rho_b}$$

$$V_a = \frac{V(\rho_a - \rho_b - 2\rho)}{\rho_a - \rho_b}$$

References

1. Wills, B.A., and Napier-Munn, T.J. (2006): *Mineral Processing Technology*, 7th Edition. Amsterdam: Elsevier Science and Technology Books.
2. 911 Metallurgist (2022a): *Dense Medium Separation HMS/DMS Process* (www.911 metallurgist.com/blog/dense-heavy-medium-separation-hms-dms, Accessed 5th May, 2022).
3. Anon (2013): *Students' Experiential Learning*. Vanderbijlpark, South Africa: Vaal University of Technology.
4. 911 Metallurgist (2022b): *Dense Medium Cyclone* (www.911metallurgist.com/equip-ment/wp-content/uploads/2018/05/dense-medium-cyclone.jpg, Accessed 5th May, 2022).
5. University of Alaska (UA) (2020): *Mining Mill Operator Training: Lesson 2: Classifying Cyclones* (https://millops.community.uaf.edu/amit-145/, Accessed 12th August, 2020).

13 Froth Flotation

13.1 Introduction

Froth flotation of ores involves the recovery of mineral values from ore slurry with the aid of chemically generated froth. The process was patented in 1906. It is the most important and the most versatile of all mineral processing techniques because it can be used to treat almost all ore minerals. The technique is versatile because it can be readily manipulated for selectivity to treat all minerals. Froth flotation is used in mineral processing, wastewater treatment and paper recycling. The differences in hydrophobicity between the mineral values and gangue minerals are enhanced with the use of surface active and wetting agents.

The froth flotation technology is applicable to many sulphides, oxides, carbonates, coal and phosphates ores. The technique can treat low-grade ores and this becomes important as the grades of ore deposits gradually decline worldwide. For instance, in the United States before 1907, the underground vein copper deposits assayed 2.5% on average, while by 1991 the average assay of copper ore mined from the deposits had decreased to only 0.6%. In the industrial treatment of wastewater, froth flotation is applied to remove fats, grease, oil and suspended solids. In paper recycling, it is used to remove printing inks and stickies. The separation principally depends on the surface hydrophobicity of mineral particles, but particle size and density have significant effects on the efficiency of the separation.[1, 2]

The reasons why the froth flotation technique is very important are as follows:

a. It permits the mining and processing of low-grade ores that would otherwise be considered un-economical to treat by other methods.
b. It allows the treatment of some complex ores because it is a selective process, that is, it can be designed to recover a particular mineral value while neglecting others. A complex ore is an ore that contains more than one useful mineral value.
c. It can also be used to treat gravity concentrator tailings that would otherwise be impossible to treat by other methods.
d. The process is selective and thus can be designed through the choice of reagents to achieve separation of specific target mineral values from various ores such as sulphides (e.g., Cu, Pb and Zn), platinum group metals (PGMs), nickel and gold-bearing sulphides, oxides such as hematite, cassiterite, minerals in an oxidized state like malachite ($Cu_2CuO_3(OH)_2$), cerussite ($PbCO_3$), nonmetallic ores such as fluorite mineral CaF_2, phosphate and fine coals. Figure 13.1 shows a flotation equipment at NMDC, Jos, Nigeria.[3]

DOI: 10.1201/9781003323433-13

FIGURE 13.1   A flotation equipment at NMDC, Jos

13.2   The General Principle of Froth Flotation

The forces of surface tension are used to float off mineral value particles from finely pulverized ore to the surface of the tank. From the mineralized froth on the surface of the water in the tank, the mineral value concentrate is scooped off while the gangue minerals settle in the tank bottom. To float on air bubbles, a target particle has to be hydrophobic, that is, water repellant or air avid or not surface wettable so that it can be supported by surface tension at the air–water interface against its weight. It is known that the majority of hydrophobic substances such as grease and waxes are covalently bonded, that is, they are nonpolar. On the other hand, metallic minerals are mainly ionic, that is, polar and because of this, they get electrostatically attracted to the bipolar water molecule. They are thus hydrophilic, that is, water "wettable".

13.3   Collectors

To float a target mineral out of the slurry, it is necessary to have a reagent, called a collector that has the capacity to alter mineral value particles surfaces to be non-wettable so that they can attach to the frother-enhanced air bubbles while being repelled by water. The collector added to the slurry will adsorb on the target mineral particles and form on them a non-wettable layer so that such particles get attached to the air bubbles and then float into a froth on the surface when the water to which a frother has been added is vigorously agitated. The collectors are organic polar/

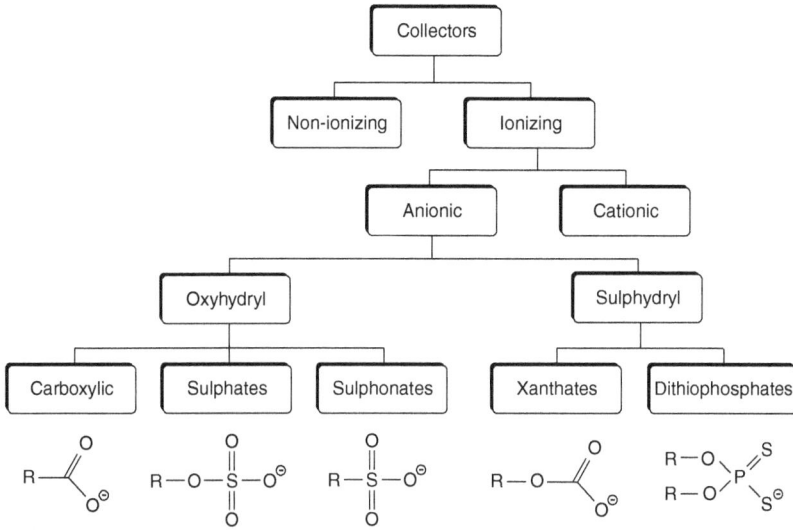

FIGURE 13.2 Classification of collectors.

Source: **Courtesy of Wikimedia (2022a)**

non-polar molecules with one end of such a molecule being ionic by which it is attached to the surfaces of the mineral value particles. On the other end is a saturated, hydrophobic, non-polar hydrocarbon covalent substance such as paraffin that is water repellant.

Collectors can be categorized as shown in Figure 13.2[4] into two: ionizing and non-ionizing collectors. Ionizing collectors dissolve in water and dissociate while non-ionizing collectors are practically insoluble in water and impart the hydrophobic property by forming a thin film on the mineral particles. The ionizing collectors are classified as either cationic or anionic depending on the charge of the water-repellant non-polar group. For example, in sodium oleate, the non-polar group is anionic while the polar group is cationic and it is thus an anionic collector. Amphoteric collectors can exhibit either cationic or anionic behaviour depending on the working pH. Collectors are added at the "starvation level" to reduce cost and to minimize the probability of floating other minerals not desired to be floated. High concentrations of collectors may also lead to multiple layers formed on the mineral value particles and this may reduce the proportion of the non-polar molecules freely oriented into the water for water repulsion. The use of longer chain collectors rather than increasing concentrations of shorter chain collectors will produce increased water repulsion but with decreased solubility of the collector in water. It has been reported by Wills and Napier-Munn (2006) that increasing the chain length decreases the collector's solubility and thus collectors in use for flotation are restricted to those having two to five carbon atoms. Furthermore, collectors with branch chains exhibit greater solubility than those with straight chains.

The ionizing collectors can be subdivided into two: oxyhydryl and sulphydryl. The oxyhydryl are based on organic, sulphate and sulphonate compounds, while the sulphydryl are compounds based on bivalent sulphur, that is, xanthates and dithiophosphates. The most commonly used organic oxyhydryl are typically organic acids, organic acid salts or soaps. An example is sodium oleate, the salt of oleic acid. Sulphonic acid with the general formula R-S (=O)2-OH where R is an organic alkali or aryl group and the R-S (=O)2-OH group a sulfonyl hydroxide is a member of the class of organosulphur compounds. Xanthates are any class of organic salts formed by treating alcohol with carbon disulphide in the presence of an alkali. Dithiophosphates (HO$_2$PS2 2) is a phosphorous oxoanion obtained by selective de-protonation of the SH group and one of the OH groups of dithiophosphoric acid. Xanthates are particularly effective for treating or concentrating sulphide ores.

The oxyhydryl sulphates and sulphonates collectors have been reported by Wills and Napier-Munn (2006)[1] to have greater selectivity but lower collecting power in comparison with the fatty acid carboxylic collectors and are thus used to float ores such as chromites, apatites, barites, mica, cassiterite and kyanite. However, the sulphydryl collectors with bivalent sulphur (thio compounds) are more widely used than the oxyhydryl collectors. The mercaptans, also called thiols, are the simplest thio compounds with the general formula RS–Na or RS–K. The xanthates (xanthogens) and dithiophosphates (aerofloat collectors) are the thiol collectors most commonly used.

Alcohol xanthates are produced when carbon disulphide reacts with simple and complex alcohols. The typical reaction is given by:

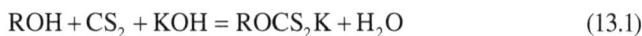

$$ROH + CS_2 + KOH = ROCS_2K + H_2O \qquad (13.1)$$

13.4   The Operating Technique

The first step is to crush and grind the ore so that its various mineral components become physically separated units as much as possible. This process is called liberation and for flotation, mineral particles should be less than 300 μm in size and can be as small as 7–10 μm. It has been noted that the liberation size of ore deposits decreases over the mining time as the near-surface ore bodies with coarse mineral grains get depleted and only ore bodies with fine dissemination previously considered difficult to treat have to be used.

The ore may be ground wet to form a slurry or the dry ore is mixed with water to form a slurry and then the mineral value desired is rendered hydrophobic by the addition of the appropriate collector surfactant. The slurry is then aerated through the hollow shaft to produce air bubbles. The impeller and diffuser enable the thorough mixing of the slurry and the air blown in. The naturally hydrophobic particles or those rendered hydrophobic by the collector attach to the air bubbles and rise to the surface to form a froth. Frothers are added to the slurry to promote the formation of a stable froth on top of the slurry. The froth is scooped with paddles from the surface to produce a concentrate.

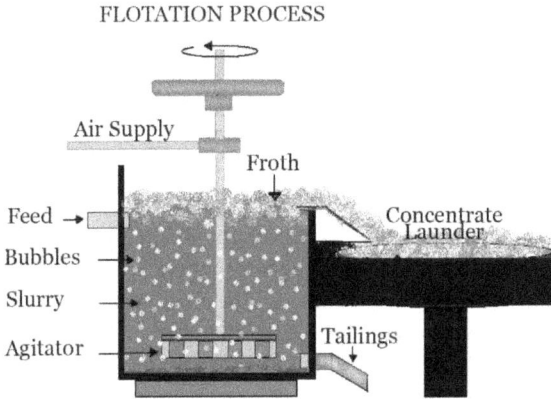

FIGURE 13.3 Froth flotation process for the concentration of a sulphide ore.

Source: **Courtesy of 911 Metallurgist (2022a)**

The minerals that remain in the bottom of the cell after flotation are called flotation tailings and may undergo further stages of flotation to obtain more mineral value particles that did not float in the first stage. This secondary treatment is known as scavenging. The final tailings are usually disposed to fill mine pits or into tailings disposal facilities for storage on a long-term basis. In a conventional mechanically agitated cell, the void fraction, that is, the volume occupied by air bubbles is low, typically between 5 and 10 per cent and the bubble size is usually greater than 1 mm. This results in a relatively low interfacial area and a small probability of contact between particle and bubble. In view of this, it becomes necessary to increase the ore particles' residence time by placing several cells in series instead of using a single large one to raise the probability of contact between particles and air bubbles. In order to maximize the desired mineral(s) recovery and to concentrate the mineral values in the concentrate, froth flotation is usually done in several stages. Figure 13.3 shows the froth flotation process for the concentration of a sulphide ore.[5]

13.5 The Flotation Stages

The flotation circuit may be a simple circuit having a single-stage rougher flotation with no provision to clean the froth as for coal flotation and a circuit with a single-stage rougher flotation, two froth cleaning stages, with no regrind. A complex circuit may consist of two stages of roughing, one stage of scavenging, three stages of cleaning, a cleaner scavenger and a regrind.

The common stages in froth flotation are:

13.5.1 Roughing

The first stage in flotation is called roughing and its product is called a rougher concentrate. The purpose of the roughing flotation is to maximize the recovery of the mineral value at as coarse a particle size as practicable with little emphasis on the

grade of the concentrate. For roughing, complete liberation is not required but liberation is sufficient to remove enough gangue minerals from the mineral value for high
recovery. A pre-flotation step to float off readily floatable materials such as organic
carbon is sometimes carried out before the roughing stage.

13.5.2   Cleaning

In the cleaning stage, the rougher concentrate for which high recovery was the focus
is normally taken for further flotation stages to remove more of the gangue minerals
that are reported to the froth. The product of the cleaning is called a cleaner concentrate or the final concentrate. The purpose of the cleaning is to obtain a concentrate
with a grade that is as high as possible. The regrinding of the rougher concentrate
is usually carried out to obtain more complete liberation before the cleaning treatments. There is usually also a re-cleaning stage with cleaner concentrates as input.

13.5.3   Scavenging

The rougher flotation tailings are often subjected to a further flotation step called
scavenging. The aim of further treatment of the tailings is to obtain any of the mineral
value targets that the initial roughing stage flotation could not recover. The recovery of additional values might become possible due to some secondary grinding to
attain further liberation of mineral values or by adopting flotation conditions that are
more intense than was used for the roughing stage. The rougher scavengers concentrates could be returned as rougher feed for fresh roughing flotation or sent to special
cleaner cells. Similarly, a scavenging step may be carried out on the cleaner tailings
from the cleaning flotation step. Figure 13.4 shows the stages in froth flotation.[6]

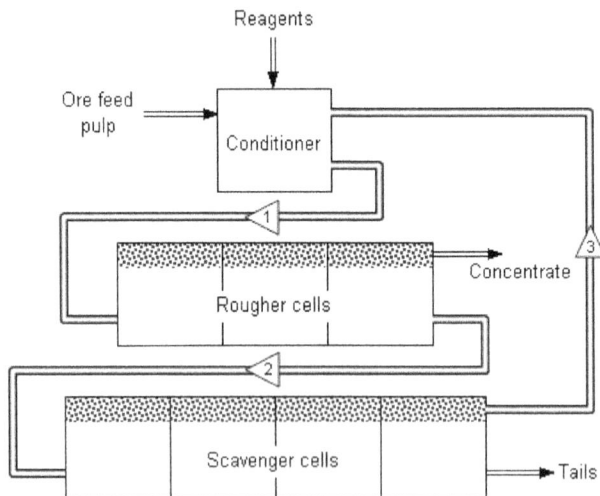

FIGURE 13.4   Stages in froth flotation.

Source: **Courtesy of Wikimedia (2022b)**

13.6   The Science of Flotation

In order to effectively float mineral values from an ore slurry, the collector used should select desired particles for wetting to prepare them for floating. The chemical or physical adsorption of the collector on the mineral value must occur to obtain the thermodynamic requirement for the particles to attach to a bubble's surface. The contact angle that the liquid/air bubble interface makes with a particle gives a quantitative measure of the wetting activity of a collector surfactant on the particle. The induction time, which is the time required for the particle and bubble to rupture the thin film separating them is another important measure for the bubbles' attachment to a particle. The rupturing is obtained by the surface forces acting between the particle and the bubble.

The bubble-particle attachment mechanism involves three steps, namely, collision, attachment and detachment. The collision occurred when particles are within the collision tube of a bubble and the velocity of the bubble and its radius determine this. The attachment of the particle to the bubble on the other hand is dependent on the induction time of the particle and bubble. The particle needs to attach to the bubble and this becomes possible if the time the particle and bubble are in contact with each other is longer than the induction time required. The induction time depends on the carrier fluid viscosity, the sizes of particles and bubbles as well as the forces between the particle and the bubbles. The detachment of a particle and bubble occurs when the shear and gravitational forces exceed the force exerted by the surface tension between them. There are cases of fine particle entrainment as these particles experience low collision efficiencies as well as sliming and degradation of the particle surfaces. On the other hand, valuable mineral recovery from coarse particles is low due to a low degree of liberation and detachment potencies between the particle and bubble that is high.

13.7   Theory of Flotation

In a slurry, froth flotation depends on the adhesion of air bubbles to selected mineral particle surfaces. The attachment of the bubbles to the surface is determined by the interfacial energies between the solid, liquid, and air phases.

Wills and Napier-Munn (2006)[1] show that:

$$W_{s/a} = \gamma_{w/a} + \gamma_{s/w} - \gamma_{s/a} \text{ and this finally leads to}$$
$$W_{s/a} = \gamma_{w/a}\left(1 - cos\theta\right) \tag{13.2}$$

where
$\gamma_{s/a}$ is the surface tension between the solid particles and air bubbles
$\gamma_{s/w}$ is surface tension between the solid particle surface and water
$\gamma_{w/a}$ is the surface tension between the water and air bubble
$W_{s/a}$ = work of adhesion required to break the bond between particle solid
     surface and air bubble interface

Due to the strong surface tension forces between the solid particle surface and the air bubbles a contact angle θ is developed. The magnitude of the contact angle θ depends on the nature of the solid surface in question. It can be seen from Equation 13.2 that as θ increases, $W_{s/a}$ increases because cos θ decreases. It can be deduced that the higher the contact angle θ the higher the Ws/a needed and the more the resistance of the system is to the forces that disrupt. The hydrophobicity of a mineral surface, therefore, becomes higher as the contact angle increases. Minerals with high contact angles θ are aerophilic or hydrophobic, that is, they have high affinity for air than for water. Most minerals are not water-repellant naturally. Collectors are therefore added and they adsorb on the mineral surfaces to render them hydrophobic and thus facilitating bubble attachment. The efficiency of a froth flotation process is denoted by a quantity called the flotation recovery, R, given by:

$$R = \frac{N_c}{\left(\dfrac{\pi}{4}\right)\left(d_p + d_b\right)^2 Hc} \tag{13.3}$$

where
   $N_c = PN_c^i$ = the product of the probability of a particle being collected (P)
      and the number of possible particle collisions $\left(N_c^i\right)$
   d_p = the diameter of the particle
   d_b = the diameter of the bubble
   H = specific height at which the recovery was calculated within the flotation cell
   c = concentration in the slurry of the particle type

13.8   Polar and Non-polar Minerals

There are two types of minerals, polar and non-polar minerals. The non-polar minerals have the following characteristics:

   a. The non-polar minerals have no net polarity
   b. The molecular bonds in them are relatively weak
   c. Non-polar minerals have covalent molecules bound together by van der Waal forces

In view of the foregoing, non-polar minerals do not readily react with water molecules. They are thus hydrophobic and are therefore naturally floatable. They have contact angles in the range of 60–90°C. They can be floated without the use of collectors but it is common to use collectors in floating them in order to increase their hydrophobicity. Examples are graphite, coal, diamond, talc and molybdenite. Wills and Napier-Munn (2006)[1] reported the grease tabling of a diamond because of its hydrophobicity.

13.9   Grease Tabling of Diamond

In grease tabling, inclined vibrating tables are covered with water-repellent grease over which the diamond ore slurry pre-concentrate is passed. The hydrophobic diamond particles become embedded in the water-repellent grease while the hydrophilic or water-seeking gangue minerals are carried along with water away from the table. The grease is then periodically or continuously skimmed away from the table and collected in a pot with holes. The pot is afterwards dipped in boiling water where the grease melts and escapes through the holes to be collected. The diamond ore particles are retained in the pot and finally taken for sorting.

13.10   Polar Minerals

Polar minerals have net polarity. They are made up of ionic or strongly covalent molecules and the polar surfaces exhibit high surface energies. The polar surface strongly reacts with water molecules and polar minerals are therefore water-seeking. Considering the degree of polarity, Wills and Napier-Munn (2006) reported that polar minerals are subdivided into five groups. Minerals such as carbonates and sulphates with ionic bonding are more strongly polar than minerals with covalent bonds. In general, Wills and Napier-Munn (2006)[1] stated that the degree of polarity increases from sulphides to sulphates, carbonates, halites, oxides-hydroxides and finally to silicates and quartz. This implies that quartz is the most polar mineral and thus the most hydrophilic.

13.11   Frothers, Activators and Depressants

The addition of a frothing agent such as pine oil or cresylic acid to the slurry is generally required. The frothing agent concentrates at the water–air interface to lower the interfacial surface tension to enable the formation of many small stable air bubbles that yield a large volume of foam. Activators and depressants may also be added as regulators for the process. The adsorption of the regulators on the surface of the mineral particles causes a reduction or increase in the attraction of the collector. The use of the regulators enables the separation of more than one mineral value and promotes a separation that is more complete and selective. The flotation of a complex ore of mixed lead, zinc and iron sulphide ores involves the use of a collector, an activator and a depressant. In the process, the addition of sodium cyanide causes a depression of the effect of the xanthate collector first added on the zinc and iron sulphides in the slurry and thus only the lead sulphide component is floated. Copper sulphate is then added to the slurry as an activator. It activates zinc sulphide by forming copper sulphide on it as a thin layer and thus particles of zinc sulphide are separated from the iron sulphide.[7]

13.12   Mechanisms of Froth Flotation

Phases in a flotation cell or vessel are the solid ore particles, the liquid carrier water and the froth. The mechanisms of froth flotation for the recovery of particles into the mineralized froth are:

a. True flotation
   Particle recovery due to chemically induced attachment to the air bubbles as desired. This is called true flotation.

b. Entrainment
   Particles recovered contrary to expectation through entrainment as water
   passes through the froth. The froth consists of many froth bubbles.
c. Entrapment
   Particles physically entrapped between particles in the froth.

The true flotation involving the attachment of mineral values to air bubbles accounts
for the major proportion of the particles in the mineralized froth concentrate. Thus,
it is the most dominant of the three mechanisms. However, the separation efficiency
of the mineral value from the gangue mineral is also dependent on the degree of
entrainment and physical entrapment of the particles into the mineralized froth.
This is because the entrainment and entrapment can recover bath mineral values and
gangue minerals, unlike the chemically induced true flotation mechanism.

In industrial froth flotation, entrainment and entrapment of gangue particles are
common and hence more than one stage of flotation may be necessary to attain the
level of mineral value content in the final concentrate product that is economically
acceptable.

For true flotation, after treatment with the collector reagents, the physico-chemical
surface property difference between the mineral-value particles and the gangue mineral particles becomes more apparent or enhanced. The mineral value particles are rendered hydrophobic and hence aerophilic, while the gangue mineral value is rendered
hydrophilic and hence aerophobic. The frothers are then added to the slurry to stabilize
the air bubbles which then attach themselves to an aerophilic particle or hydrophobic
particle and are after lifted to the water surface to form the mineralized froth.

The impeller agitator is used to make the slurry sufficiently turbulent in order to
promote the rate at which particles and air bubbles collide resulting in the mineral
value particles attaching to the air bubbles and their subsequent movement to the
mineralized froth for recovery. The froth flotation process is suitable to recover fine-
sized particles with sizes less than 300 µm. Large-sized particles cannot be floated
because the adhesion between them and air bubbles will be less than the large particle's weight. A carrier bubble will thus have to drop the particle load. For successful
froth flotation, an optimum particle size range is therefore required.[1]

There are two types of flotation, direct and in-direct flotations. In direct flotation,
the mineral value particles are rendered hydrophobic and transferred to the mineralized froth phase, while in indirect or reverse froth flotation; the gangue mineral
particles are found more convenient to be rendered hydrophobic and then transferred
to the surface of the slurry. The overall selectivity of the froth flotation process is
improved by the frother phase derived from the frother reagent addition. The function of the froth phase are:

a. The frother that generates the froth phase stabilizes the air bubbles to attach
   to particles and also keep them from bursting
b. It also reduces the recovery of the water-entrained particles and preferen-
   tially retains the attached mineral value particles

Figure 13.5 shows the flow of slurry in industrial flotation.[8]

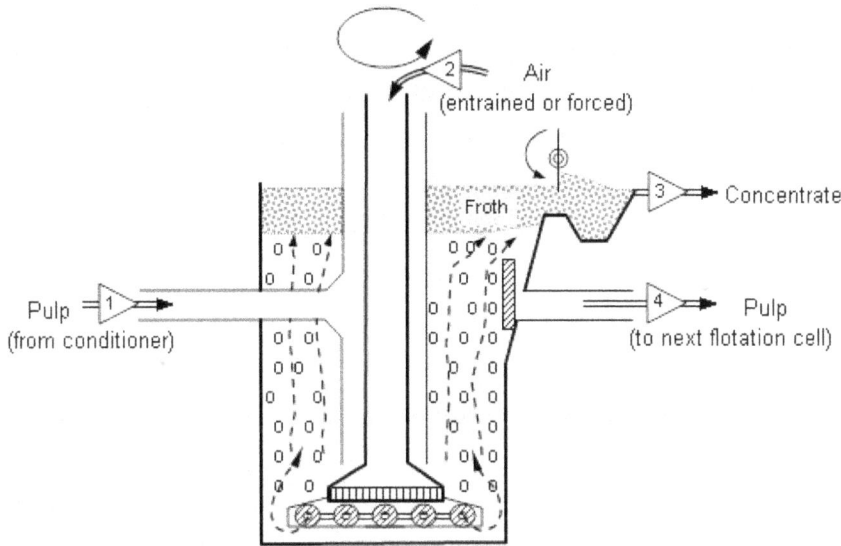

FIGURE 13.5   The flow of slurry in industrial flotation.

Source: **Courtesy of Wikimedia (2022c)**

13.13   Column Flotation

Flotation column was invented in 1962. It was based on the earlier work on solvent extraction at the Eldorado Mining research laboratory. The column as used today was earlier applied to carry out solvent-in-pulp processing of uranium ore. The ore slurry was fed near the top of a laboratory column and at the bottom solvent droplets are generated. The density difference leads to a countercurrent flow of the slurry particles and the solvent which causes a solvent phase with a distinct pulp–surface interface to be developed at the column's surface. An aqueous diluent is then introduced near the interface to reduce solvent contamination by the slurry particles. Slurry is drawn out of the column bottom, while the solvent overflow from the top lip of the column. Boutin and Tremblay applied the principle described to ore flotation by substituting the diluent with water and the solvent droplets with air bubbles.

The method was first successfully applied to amine flotation of silica from iron ore samples. The Column Flotation Company of Canada carried out the first notable successful industrial column flotation on molybdenum cleaning. It is now used in the cleaning of sulphides of copper, zinc, lead and molybdenum as well as for the flotation of coal, phosphate and iron ore. The marked difference between the conventional mechanical and column flotation are:

1. There is no mechanical agitation/shear in column flotation
2. The column cell is relatively tall and narrow in comparison to mechanical flotation vessels
3. The froths in column cells are deeper and wash water is liberally used on its froth

Water Sprays

Froth Zone

Feed Inlet

Column Tank

Collection Zone

Spargers

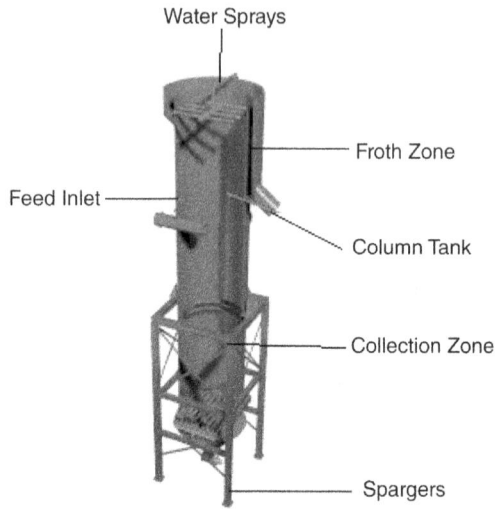

FIGURE 13.6   Schematic diagram of a column flotation cell.

Source: **Courtesy of 911 Metallurgist (2022c)**

In column flotation, the turbulence due to the ascending air bubbles causes the mixing of the slurry as against the impeller used in the mechanical cells. Column flotation is used to float ores that have been ground such that the mineral values are sufficiently liberated. The technique produces concentrates of higher grades at lower energy consumption in comparison with conventional mechanical cells.

Columns are mostly used to produce final-grade concentrates because they are highly selective. Columns also differ from mechanical cells in terms of their shape, longer froth depth, the system of generating bubbles and the use of wash water. The wash water is required to stabilize the froth by replacing the water draining down by gravity into the collection zone. The sparger's air rate determines the position of the pulp/froth interface and influences the stability of the column. When the air rate decreases, the position of the interface will be raised and thus lowering froth depth and vice versa. The air rate has to be maintained below a critical rate to ensure the stability of the column. Column cells are typically designed to nave heights varying from 5 to 15 m and diameter of between 1 and 3 m, while the froth depth ranges from 50 to 200 cm. They can be square, circular or rectangular in cross-section. Figure 13.6 presents a schematic diagram of a column flotation cell, while Figure 13.7 shows an Eriez column flotation machine, while Figure 13.7 presents Eriez column flotation machine.[9,10,11]

13.14   Flotation Conditioning

Flotation conditioning is carried out to ensure even dispersion of the reagents added at starvation levels in the slurry, its intimate admixing with the pulp particles multiple times and provision of adequate air-bell contacts to stimulate flocculation of the

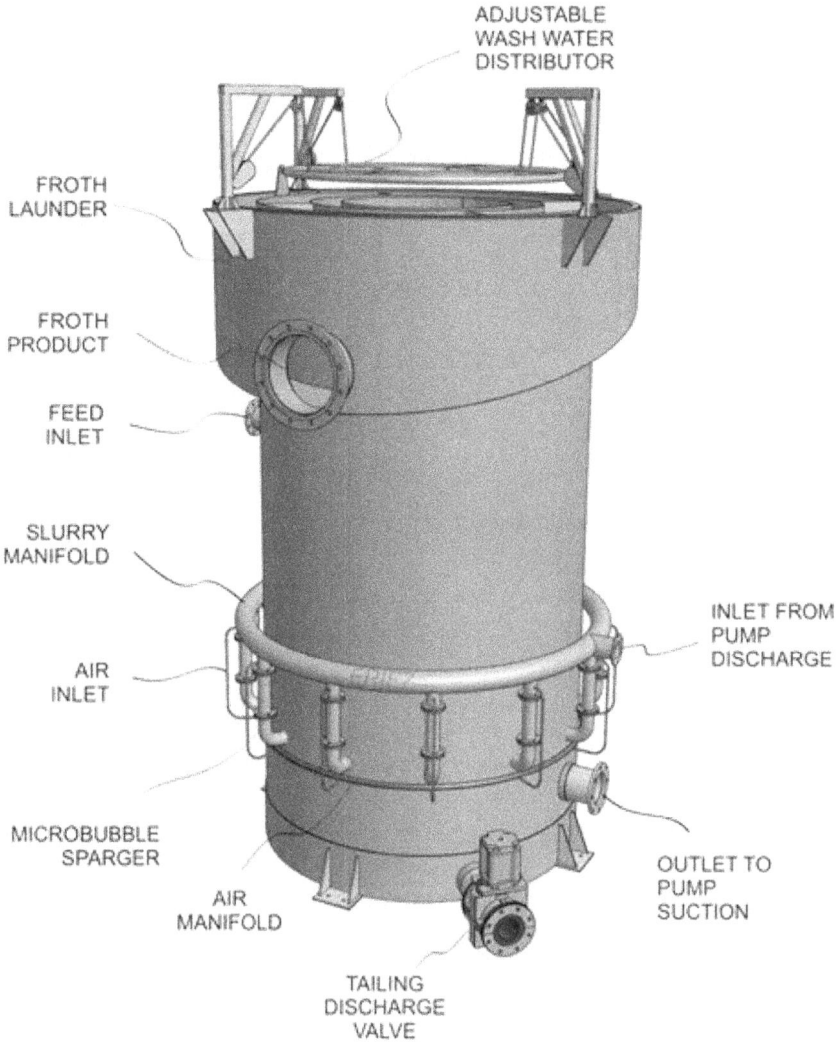

FIGURE 13.7   Eriez column flotation machine.

Source: **Courtesy of Eriez (2022a)**

target minerals. The duration of the conditioning may be a few minutes if carried out in an intensely agitated mechanical operated cell having an impeller, while a longer conditioning time is required in a less intensely stirred conditioning tank for discharge into pneumatically operated cells with no agitation.

The importance of adequate conditioning allowed in the conditioning tank was observed in the use of a fatty acid collector to float silica. The conditioning tank can also be equipped with an automatic water addition valve to control slurry density that affects the succeeding flotation process based on feedback from the tank exit pulp density measurement or a tank-installed differential pressure sensor. It can

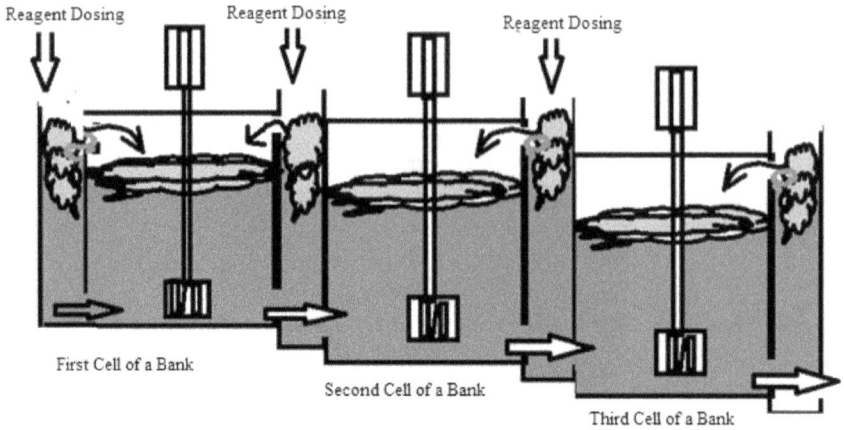

FIGURE 13.8 Flotation conditioning in a bank of cells.

Source: **Courtesy of 911 Metallurgist (2022c)**

also serve as a buffer to prevent a turbulent flow of slurry into the first cell making support of a stable froth impossible. Figure 13.8 shows a flotation conditioning in a bank of cells.[12]

A typical flotation bank is designated as F-4-I-3-D, that is, the bank comprises a feed box, four cells, an intermediate box, three cells and a discharge box. For industrial froth flotation, the cell parameters have been standardized. For example, to float a copper ore, the % solids, retention time and the number of cells are 25–32, 8–10 min and 8–12, respectively. For a phosphate ore, the parameters are 30–3, 4–6 mins and 4–5 mins, for % solids, retention time and the number of cells, respectively.[13]

Factors that Affect the Efficiency of Froth Flotation

The factors that influence the performance of an industrial froth flotation process include the following.

13.14.1 Air Control

There must be a balanced airflow into the cell. If the airflow is too low, it will not create enough bubbles to carry the concentrated particles into the launder and thus lowering recovery. The required air flow rate must therefore be maintained.

13.14.2 Level Control

The slurry level in the flotation cell must be kept at the set level. If the slurry level is too much below the set level, the concentrate froth will not be able to flow into the launder and if the level is too high above the set level, the froth and part of the slurry will be delivered into the launder.

13.14.3   Feed Rate

Flotation cells are designed to receive slurry feed at a specified rate. A delivery above the required range will reduce the retention time of particles in the cell and this may lead to a loss of valuable minerals that would be recovered.

13.14.4   Overdosing of Reagents

Overdosing of the collector can lead to mineral value particles' surface becoming too hydrophobic and causing the breaking of the froth carrier.

Typical Froth Flotation Tests

13.14.5   Procedure for Laboratory Froth Flotation Test

Procedure for a typical froth flotation test is described as follows:

1. One kilogram of slurry at pulp density between 25% and 40% is delivered into the 1 kg capacity Denver cell.
2. The slurry in the cell is then agitated with the impeller at say 900 rpm or higher setting to ensure homogeneous suspension of solids for a specific time, say, 2 min. The agitation must be sufficiently high but without scattering the mineralized froth column to ensure that all the solid particles are kept in suspension.
3. Collector solution is added to the slurry and the reagent is conditioned with the slurry by agitating the slurry, for say, 3 min. Conditioning is an agitation period prior to air inlet to allow ore particles surfaces to react with the reagent and it can vary from a few seconds up to 30 min.
4. The frother reagent is afterwards added and conditioned with the slurry for say, 2 min. The froth quantity should be small and with a depth of 2–5 cm. Then the air control valve is opened to let in the air into the impeller.
5. Concentrate froth is collected every 15 s using a stopwatch by two sweeps of the paddles one immediately after the other from the back of the cell to the front.
6. Remove the froth into three containers in three steps of 3, 7 and 15 min.
7. It is important to maintain a constant pulp level of 15–20 mm below the cell lip and this is achieved by adding water as required.
8. The froth concentrate is pressure filtered and dried.

The mass pull of concentrate during batch flotation test is given by:

$$Mass\,Pull\,(\%)\,of\,Concentrate\,A = \frac{Mass\,of\,Concentrate\,A}{Total\,dry\,mass} \times 100 \qquad (13.4)$$

and so on for other subsequent concentrates.

13.15 Typical Laboratory-scale Froth Flotation Tests

Madilen et al. (2015)[14] reported the flotation of blends of middle groups seams of PGMs from the South African Bushveld Igneous Complex. The Stirred Mill Detritor (SMD) was used to bead mill the ore for 15 min. A pulp was obtained by mixing 3.6 kg of the ore mixture ground 55% passing 75 μm sieve with water in an 8-litre Denver cell. The pulp was dosed at 180 g/t with Sodium Iso-butyl with Xanthate (SIBX) collector. The conditioning of the pulp with the collector took 2 min. The Sendep 30D at 80 g/t depressant was added to the pulp and conditioned with it for 2 min, while Senfroth 38 was also added dropwise to the pulp. The batch float concentrate collection was carried out in three timed increments of concentrate collections at 5, 10, and 15 min at 15 s intervals.

Otunniyi et al. (2016)[15] reported the activation of a Botswana complex pentlandite-pyrrhotite-chalcopyrite ore with copper sulphate in a 2 L Denver cell at an impeller speed of 1200 rpm and pH of about 9. In the base case conditioning without $CuSO_4$ addition, the first treatment was done using potassium normal butyl xanthate at a dosage of 75 g/ton and at 3 min conditioning time. The other three treatments were done with copper sulphate dosages of 15, 30 and 45 g/ton, respectively, and the pulp was conditioned for 5 min before collector addition. In each flotation run, five concentrates were separately collected over the time intervals between 0, 1, 3, 6, 10 and 15 min of flotation. The fractions from every run were assayed for nickel, copper, iron and cobalt from hot aqua regia digestion using an atomic absorption spectrometer. From the assays, the mass pulls and sink values, and the feed grades were determined for every run. Cumulative grades and recoveries were computed for the various constituents to analyse the responses to the treatments. The results obtained showed an increase in nickel recovery and grade with copper sulphate dosage addition, but the trend showed a maximum after which depression was obtained. The optimum dosage in this instance was between 15 and 30 g/ton for the best grade recovery. Therefore, to achieve the activation of pentlandite in the ore with the collector, copper sulphate dosage must be at an optimum level.

Mokegthwa et al. (2016)[16] reported the construction of a milling curve for the South African Palabora carbonatitic copper ore. The ball mill drum was loaded with 1.5 kg of ore and 7.2 kg of steel balls in 750 ml of water. The slurry mixed with steel balls was then milled for 10 min. The grinding balls were removed and the ground ore was emptied into a bucket. The milled sample was wet screened through a 75 μm sieve. The wet-screened sample residue was dried and weighed to determine the percentage passing the 75 μm sieve. The procedure described was repeated at 20, 30, 40, 50 and 60 min grinding times. The tests were carried out in duplicates and the data obtained was used to construct a milling curve.

Mokegthwa et al. (2016)[16] also reported the froth flotation concentration of the Palabora carbonatitic copper ore with sodium ethyl xanthate collector prepared in water at 60 g/t strength. The prepared solution was thoroughly mixed for 5 min on the magnetic stirrer hotplate using a magnetic bar set to stir at 400 rpm at 25°C and the lime solution was also prepared to a pH of 11. About 1.5 kg milled sample was poured into the 4.5-litre flotation cell and water was added to make up the level of the cell. The flotation cell was placed in the base of the Denver flotation machine.

The flotation impeller was lowered into the slurry and the concentrate collection pan was placed under the overflow lip of the cell. The agitator was started and the impeller was set to 1000 rpm. Lime and sodium ethyl xanthate collector were added and each was conditioned with the slurry for 2 min. Methyl isobutyl ketone was added as the frother. The air valve was opened to produce the concentrate frothing. The froth was then scraped across the top of the cell towards the overflow every 15 s in the 3, 7 and 10 min scraping stages.

13.16 Pilot-scale Froth Flotation Test

Ola et al. (2009)[17] reported the pilot-scale froth flotation of Itakpe sinter grade concentrates to produce a super-concentrate suitable for Midrex direct reduction process as shown in Figure 13.9 The concentrate was taken into the ball mill at a rate of 500 kg/h. The pilot plant flotation operation was carried out sequentially in a bank of flotation cells of 130 litres capacity each. The iron ore was treated with 110 g/ton of alkyl ether amine Flotigam EDA as a collector and 10 g/ton of Flotanol M as a frothing agent. The slurry's pH was adjusted to 10.5 by the addition of sodium hydroxide and it was further treated with maize starch causticized with 25% NaOH as an iron oxide depressant. The slurry at 50% solids was then conditioned in the bank of 4 roughing flotation cells for 2 min. The slurry now 40% solid was de-slimed at 5 µm size using 2 mm hydrocyclones operating at 50 Psi pressure. The cyclone underflow having 33% solids was delivered to the 4 cleaner cells and cleaned for 2 min. The cleaned slurry at 33% solids was re-cleaned in the bank of 2 re-cleaning cells for another 2 min. Figure 13.9 shows the flowsheet for superconcentrated production from Itakpe iron ore.[16]

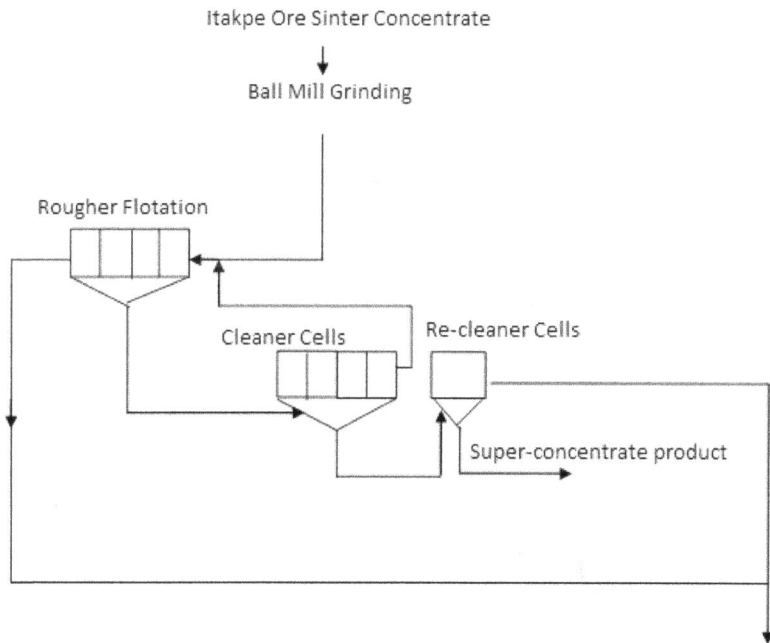

FIGURE 13.9 Flowsheet for superconcentrated production from Itakpe iron ore (Ola et al., 2009).

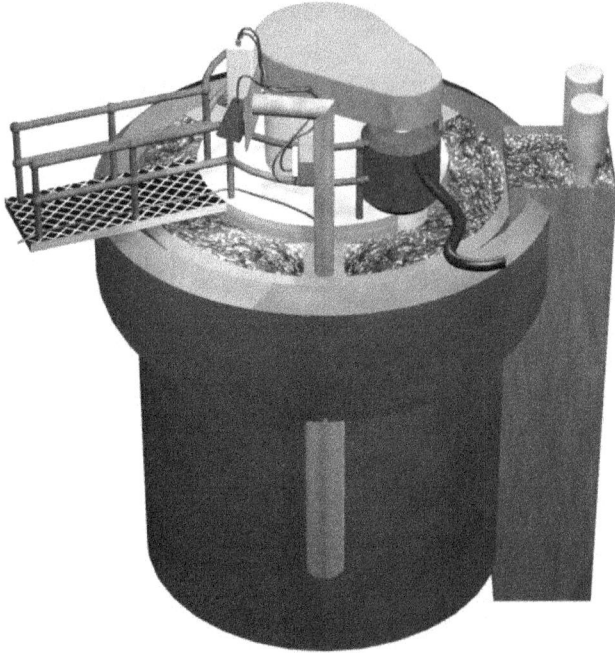

FIGURE 13.10 Diagram of a cylindrical flotation cell with camera and light used in image analysis of the froth surface.

Source: **Courtesy of Wikimedia (2022d)**

FIGURE 13.11 An Eriez industrial froth flotation plant.

Source: **Courtesy of Eriez**

It has been established that a combination of the various conventional mineral processing techniques is required for the economic treatment of ore. For instance, the cheap but less selective gravity method is often used to pre-concentrate ores by using the specific gravity property to reject a major part of the less dense gangue contents while the more expensive and more selective techniques like flotation are used to efficiently produce acceptable final clean concentrates. Figure 13.9 shows the flowsheet for a superconcentrate production from Itakpe iron ore, Figure 13.10 presents a diagram of a cylindrical flotation cell, while Figure 13.11 shows an Eriez industrial froth flotation plant.[18–20]

References

1. Wills, B.A., and Napier-Munn, T.J. (2006): *Mineral Processing Technology*, 7th Edition. Amsterdam: Elsevier Science and Technology Books.
2. Wikiwand (www.wikiwand.com/en/Froth_flotation, Accessed 11th July, 2020).
3. Ayodele, T.J. (2020): *Optimization of Hydro Extraction of Copper from a Hematite Dominated Copper Ore*. PhD Qualifying Examination Report, Department of Materials Science and Engineering, Obafemi Awolowo University, Ile Ife, Nigeria Discourse (https://ask.learncbse.in/t/draw-the-diagram-showing-i-froth-floatation-ii-magnetic-separation/10062, Accessed 13th July, 2020).
4. Wikimedia (2022a): *Classification of Collectors* (https://upload.wikimedia.org/wikipedia/commons/8/83/Collectors.png, Accessed 23rd May, 2022) By EricaBilodeau—Own work, CC BY-SA 4.0 (https://commons.wikimedia.org/w/index.php?curid=40839729).
5. 911 Metallurgist (2022a): *Flotation Separators* (https://img-proxy.blog-video.jp/images?url=http:%2F%2Fwww.911metallurgist.com%2Fblog%2Fwp-content%2Fuploads%2F2013%2F09%2Fflotation-separators.jpg, Accessed 1st April, 2022).
6. Wikimedia (2022b): *Stages in Froth Flotation* (https://upload.wikimedia.org/wikipedia/commons/5/55/FlCirc.PNG) By Thermbal at English Wikipedia.—Own work, Public Domain (https://commons.wikimedia.org/w/index.php?curid=28001668, Accessed 23rd May, 2022).
7. Cottrel, A. (1980): *Introduction to Metallurgy*. London: ELBS Books.
8. Wikimedia (2022c): *The Flow of Slurry in Industrial Flotation* (https://upload.wikimedia.org/wikipedia/commons/1/19/FlCell.PNG).
9. University of Alaska, Fairbank (UAF) (2020): *Mining Mill Operator Training: Lesson 2: Classifying Cyclones* (https://millops.community.uaf.edu/amit-145/, Accessed 12th August, 2020).
10. Eriez (2022a): *Eriez Column Flotation* (www.eriez.com/NA/EN/Flotation/Column-Flotation.htm#:~:text=Eriez%20Column%20Flotation%20The%20Eriez%20Flotation%20has%20supplied,energy%2C%20and%20specialty%20applications%20such%20as%20oil%2Fwater%20separation, Accessed 27th May, 2022).
11. 911 Metallurgist (2022b): *Column Flotation System* (https://z4y6y3m2.rocketcdn.me/blog/wp-content/uploads/2016/02/Column_Cells.png, Accessed 25th April, 2022).
12. 911 Metallurgist (2022c): *Flotation Conditioning* (www.911metallurgist.com/blog/flotation-conditioning).
13. Metso (2022): *Basics in Mineral Processing* (https://vdocument.in/basics-in-minerals-processingmetso.html, Accessed 23rd June, 2022).
14. Madilen, N.M., Adeleke, A.A., and Mendonidis, P. (2015): Effects of Sodium Iso-butyl Xanthate Dosage on the Froth Flotation of Bead Milled Middle Group 1-3 PGM Ore Blend. *Aceh International Journal of Science and Technology*, 4(1), 59–63.

15. Otunniyi, I.O., Oabile, M., Adeleke, A.A., and Mendonidis, P. (2016): Copper Activation Option for a Pentlandite–Pyrrhotite–Chalcopyrite Ore Flotation with Nickel Interest. *International Journal of Industrial Chemistry* (https://doi.org/10.1007/s40090-016-0087-7).

16. Mokegthwa, M.F., Adeleke, A.A., Mendonidis, P., and Adeoye, M.O. (2015): An Evaluation of Sodium Ethyl Xanthate for the Froth Flotation Upgrading of a Carbonatitic Copper Ore. *Journal of Physical Science*, 27(2), 13–21.

17. Ola, S.A., Usman, G.A., Odunaike, A.A., Kollere, S.M., Ajiboye, P.O., and Adeleke, A.O. (2009): Pilot Scale Froth Flotation Studies to Upgrade Nigerian Itakpe Sinter Grade Iron Ore to a Midrex-Grade Super-Concentrate. *Journal of Minerals and Materials Characterization and Engineering*, 8(5), 405–416.

18. Wikimedia (2022d): *Cylindrical Industrial Flotation Cell* (https://upload.wikimedia.org/wikipedia/commons/d/d0/Flotation_cell.jpg, Accessed 23rd May, 2022).

19. Wikimedia (2022e): *Diagram of a Cylindrical Flotation Cell with Camera and Light Used in Image Analysis of the Froth Surface* (https://upload.wikimedia.org/wikipedia/commons/d/d0/Flotation_cell.jpg, Accessed 23rd May, 2022) By Dhatfield—Own work, CC BY-SA 3.0 (https://commons.wikimedia.org/w/index.php?curid=4264037, Accessed 23rd May, 2022).

20. Eriez (2022b): *Eriez Flotation Division* (https://th.bing.com/th/id/OIP.IdicTWl6SLa0GjePc7ihcQHaFj?pid=ImgDet&rs=1, Accessed 5th May, 2022).

14 Magnetic Separation

14.1 Introduction

In magnetic separation, the difference in the magnetic susceptibility between the mineral particles in an ore is used to separate them. Magnetic susceptibility is the degree to which a material can be magnetized in an externally applied magnetic field or the susceptibility of a material to acquire a magnetic moment when exposed to a magnetic field or the tendency of a material to concentrate a magnetic field within itself. The separation based on magnetic susceptibility may be of the following types:

a. Separation of magnetic valuable mineral(s) in association with gangue minerals which are non-magnetic from such non-magnetic contaminants. For instance, the separation of magnetite, which is ferromagnetic from quartz which is diamagnetic.
b. Separation of a magnetic contaminant from non-magnetic values. For instance, the non-magnetic copper and tin-bearing valuable minerals which are often found in association with magnetite and wolframite, which are ferromagnetic and paramagnetic, respectively and in trace concentrations; can be concentrated further by removing the magnetite and wolframite contaminants.
c. Separation of valuable minerals which are magnetic from valuable non-magnetic minerals.

14.2 Principles of Magnetism

If the ratio between the induced magnetization within the material and the inducing field or the applied magnetic field strength, is expressed per unit volume, volume susceptibility (k), a dimensionless quantity, is defined as:

$$k = \frac{M}{H} \tag{14.1}$$

where
   M = the volume magnetization induced in a material of susceptibility k
   H = the applied external field

Magnetic susceptibility occurs by the interactions of electrons and nuclei in the atom with the externally applied magnetic field. Magnetic flux is a measurement of the total magnetic field which passes through a given area. It can be quantitatively defined as force per unit length, per unit current acting on a conductor placed at 90° to the magnetic field. The magnetic flux density indicates the strength of a magnetic

DOI: 10.1201/9781003323433-14

field. Nuclei and electrons possess a quantum mechanical property called spin and can be considered as tiny spinning magnets. Minerals have different susceptibility ranges. It has been reported that the susceptibility ranges for magnetite, hematite and cassiterite are >6505 × 10^{-9}, 720–6505 × 10^{-9} and <22.5 × 10^{-9} m^3/kg, respectively.

Volume magnetic susceptibility can be measured by determining the force change obtained upon a substance when a magnetic field gradient is applied. In the simple test, a sample is hung between the poles of an electromagnet and the Gouy balance is used to measure the force change that occurred. A simple induction device can be used instead of the Gouy balance.

When the spins of the nuclei and the electrons in a material are oriented in the same direction as an applied external magnetic field, their individual magnetic moments locally augment the field. The augmentation of the external magnetic field is called paramagnetism. Sub-atomic particles in a material may also create magnetic effects that oppose the applied magnetic field and this is called diamagnetism. Since electrons are very small in comparison to the nucleus, their spins are considered concentrated into smaller volumes. In view of this spin/size difference, electron-magnetic field interactions are much stronger than nucleus-magnetic field interactions. Therefore, electrons primarily determine the overall magnetic susceptibility of a material.

Apart from possessing a spin, electrons also orbit the nucleus and possess a second quantum property called orbital angular momentum (L). Orbitals with paired electrons promote diamagnetism, while those with unpaired electrons promote paramagnetism. In most materials, the bulk susceptibility is determined by the net sum of the competing diamagnetic and paramagnetic effects. The major mechanisms contributing to susceptibility include Langevin diamagnetism, Van Vleck paramagnetism, nuclear paramagnetism and Curie paramagnetism. In the latter, the unpaired electrons in the orbitals experience net alignment with the externally applied magnetic field and this alignment augments the applied field and therefore makes the material paramagnetic. The element gadolinium has one of the strongest Curie paramagnetic effects because it has seven unpaired electrons in its shells. Minerals can be classified as one of the ferromagnetic, paramagnetic and diamagnetic groups based on the magnetic properties exhibited.

14.3 Type of Materials Based on Magnetism

The three groups of materials based on their behaviours in the presence of an externally applied magnetic field are described as follows.

14.3.1 Ferromagnetic Materials

Ferromagnetic materials contain unpaired electrons, with each unpaired electron having a small magnetic field of its own. The magnetic fields in unpaired electrons align readily with each other in response to an applied external magnetic field. This alignment tends to continue even after the magnetic field is removed and this phenomenon is called hysteresis. In contrast to other substances, ferromagnetic materials are magnetized easily, and in strong magnetic fields, the magnetization approaches a definite limit called saturation. They also retain the induced magnetism after the removal of the external field. Examples of ferromagnetic minerals are magnetite, franklinite, maghemite, pyrrhotite, titanomagnetite and roasted hematite. Other ferromagnetic materials include iron, cobalt, nickel, and some alloys or compounds containing one or more of these elements. Ferromagnetism also occurs in gadolinium and a few other rare-earth elements.

14.3.2 Paramagnetic Materials

Paramagnetism is a type of magnetism where a material is medium or weakly attracted by an externally applied magnetic field and forms an internal, induced magnetic field in the direction of the applied magnetic field. Examples of paramagnetic minerals are hematite, siderite, wolframite, Ilmenite, limonite, chromite, fergusonite, pyrolusite, samarskite, spinel and garnet. Aluminium and titanium are examples of paramagnetic elements. Paramagnetism is also due to the presence of unpaired electrons in the material and therefore most materials having atoms with incompletely filled atomic orbitals are paramagnetic.

14.3.3 Diamagnetic Materials

The electrons in diamagnetic material occur paired together in an orbital and their total magnetic spin is zero. Atoms with diamagnetic electrons are called diamagnetic atoms. Diamagnetic atoms are repelled from a magnetic field. Examples of diamagnetic minerals are feldspar, quartz, mica, halite, calcite, amphibole, cassiterite, stibnite, vanadate, talc, topaz, tourmaline, tremolite, barite, baddeleyite, beryl, azurite, blende, chalcocite, cinnabar, diamond, fluorite, dolomite, monazite, muscovite and molybdenite. To separate ferromagnetic and paramagnetic minerals, the relative magnetic field intensity required varies and increases as the magnetic susceptibility of a material decreases, taking that of ferromagnetic materials, as unity; as shown in Table 14.1.[1–4]

TABLE 14.1
Magnetic Field Strength Requirements to Separate Minerals

Category	Solid Mineral	Magnetic Field Strength Required
Ferromagnetic minerals	Magnetite	1
	Pyrrhotite	0.5–4
Paramagnetic materials	Ilmenite	8–16
	Siderite	9–18
	Chromite	10–16
	Hematite	12–18
	Wolframite	12–18
	Tourmaline	16–20

14.4   Principle of Magnetic Separation

The principle of magnetic separation is shown in Figure 14.1.[5]

A simple magnetic separator consisting of a magnetic and a non-magnetic wheel is shown in Figure 14.1. It received a powdered ore and based on their magnetic susceptibility difference separate the particles into two.

14.5   Categories of Magnetic Separators

In a typical magnetic separator, the mineral sample is passed on a metal trough through a magnetic field emanating from an electromagnet or a permanent magnet and the charge is split based on the magnetic susceptibility of its contents near its exit end. The magnetic field's strength is varied to separate the minerals in the sample into ferro, para and diamagnetic. Magnetic separators are operated dry or wet and under low or high magnetic field intensity. For ore particles larger than 0.5 cm in diameter, dry magnetic separators are used. However, for ore particles less than 0.5 cm in diameter, wet magnetic separators are used to avoid dust generation and to attain better separation efficiency. Similarly, when the applied magnetic field strength is lower than 0.3–0.7 T, the magnetic separation is said to be low intensity and when it is higher, it is a high-intensity separation. Therefore, we can have the following magnetic separators types:

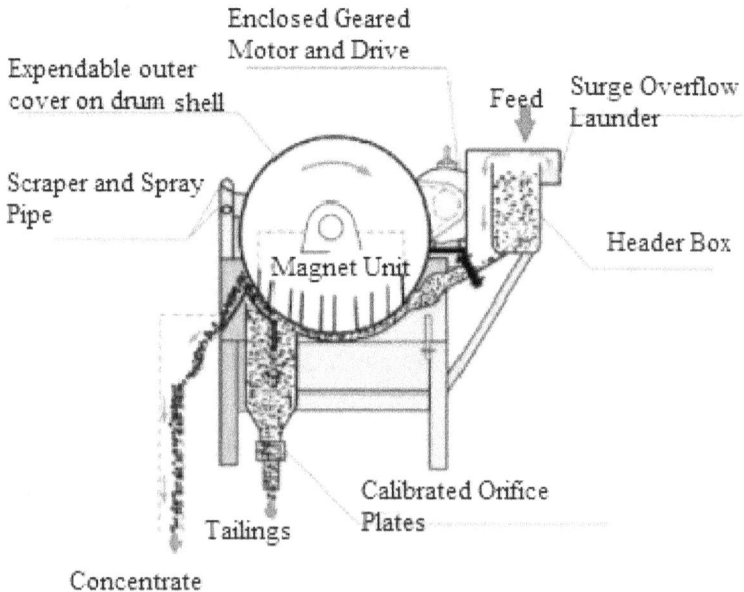

FIGURE 14.1   Principle of magnetic separation.

Source: **Courtesy of 911 Metallurgist (2022a)**

a. Low-intensity dry magnetic separators (LIDMS)
b. Low-intensity wet magnetic separators (LIWMS)
c. Dry high-intensity magnetic separators (DHIMS)
d. Wet high-intensity magnetic separators (WHIMS)

Magnetic separators of low magnetic field intensities are used to recover minerals such as magnetite and pyrrhotite which are ferromagnetic. The magnetic separations can be done dry or wet with magnetic field intensity that can be up to 2 T and can take large ore particles. Magnetic separators of high magnetic field intensity are used to recover minerals such as malachite, hematite, rutile, pyrite, bornite, monazite, siderite, chromite, ilmenite and biotite, which are paramagnetic. It can be cyclic and continuous. The separation is however, restricted to the processing of ore particles with sizes below about 1000 μm.

14.6   Design in Magnetic Separation

The basic requirement for magnetic separation is the presence of a high- or low-intensity magnetic field having a steep field strength gradient. When placed in a field of uniform flux between flat poles, particles with magnetic susceptibility will get oriented but will not change position. But in a converging field, the particles will not only get oriented but will move. The converging field is created by making one of the two poles V-shaped. The tapering of the V-pole causes a concentration of the magnetic flux unto a small area producing high intensity. The flat pole has small magnetic flux intensity per unit area, while the V-pole has a very high magnetic flux intensity over its small area. There is therefore a steep magnetic flux intensity across the gap between the flat and V poles.[1]

14.7   Wet Drum Magnetic Separation

Low-intensity separators, that is, magnetic separators that provide low-concentrating magnetic field strengths lower than about 0.3 T are used to concentrate ferromagnetic and some highly paramagnetic minerals because such minerals exhibit high magnetic susceptibility at low applied field strengths. For instance, the low-intensity drum magnetic equipped with ferrite-based magnets is known to provide a maximum magnetic field strength below 0.3 T on its surface and 0.12 T at a radius of 50 mm from the surface. Hematite and siderite are usually roasted to convert them to a ferromagnetic state and hence respond well to separation in low-intensity separators. The drum separator can be of concurrent, counter-current or counter-rotation type with typical feeds of 0–6, 0.6–6 and 0–0.5 mm, respectively.

The dry low-intensity drum magnetic separator in Figure 14.2[6] has a diameter of about 6 cm and a rotation speed of 33 revolutions per minute (rpm). The permanent magnet produced a maximum magnetic flux density of about 160 mT on the drum's surface. The operation of the separator drum under a low magnetic flux intensity means that only ferromagnetic materials can be taken out by the drum.

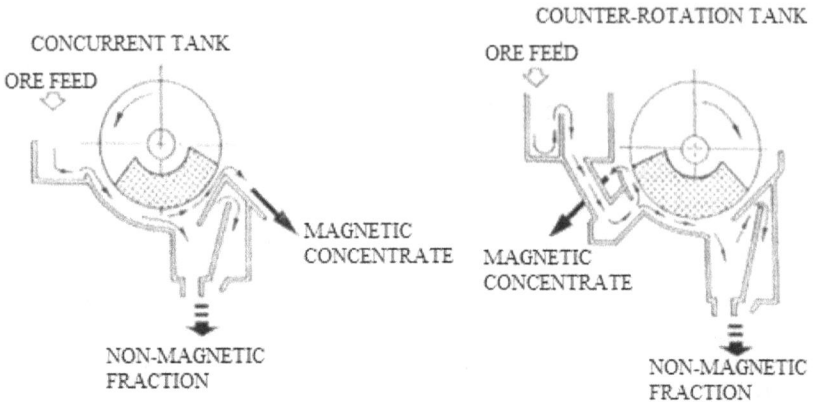

FIGURE 14.2   Dry drum magnetic separator with concurrent and counter rotation tank.

Source: Courtesy of 911 Metallurgist (2022b)

Roll Separators

Eriez' Rare Earth (RE) Roll Separators are used to treat a wide variety of coarse ore materials which do not respond well to traditional methods of processing on low-intensity dry drum separators or high-intensity-induced magnetic roll separators. It is equipped with very strong high gradient magnetic circuits having magnetic attraction about ten times that of the conventional ceramic type magnets. The RE Roll can take in ore feeds of particle sizes as coarse as 13 mm and feeds of very fine size with a varying degree of magnetic susceptibility. The separator has magnetic field strength that exceeds 2.1 T at the magnetic circuit pole. It is suitable for paramagnetic minerals and very fine-sized ferromagnetic minerals below 325 mesh.[7]

14.8   Operation of a Wet High-intensity Magnetic Separator (WHIMS)

The Eriez GZRINM wet high-intensity magnetic separator is a vertical ring, high gradient magnetic separator. It has a matrix-bearing separating compartment where a DC current is applied to the coil to produce a magnetic field. The moving ring or carousel is filled with the matrix which is a special type of magnetic iron. When the matrix is within the electromagnet's magnetic field, it becomes induced magnetically. Ore slurry is fed into the magnetized matrix and is taken through the magnetic field. During the movement, the very high magnetic gradient will attract the weakly magnetic ore particles, while the non-magnetic particles are allowed to pass through.[8] Figure 14.3 shows the Eriez GZRINM wet high-intensity magnetic separator.[7]

FIGURE 14.3   The Eriez GZRINM wet high-intensity magnetic separator.

Source: **Courtesy of Eriez (2022a)**

14.9   A Laboratory Bench Whims Practice

An oxidized taconite ore assaying 36.9% Fe was ground in a rod mill to 80% <100 mesh. One kg of rod mill discharge was then treated on the laboratory scale in an Eriel Model L-4 WHIMS with a 20 cm deep matrix of 0.2 cm iron spheres at 0.5 T applied field. The concentrate obtained assayed 51.8% Fe which represented 96.2% of the feed Fe content, while the tailings assayed 4.5% Fe and translated to 3.8% of the Fe content of the feed. The concentrate and tailings weights were 68.5% and 31.5% of the feed weight, respectively. The first concentrate, called the pre-concentrate was further ground in a ball mill to 80% <400 mesh (37 μm) and treated with a matrix of coarse stainless steel wool at 1.0 T to produce concentrate and tailings that assayed 61% and 24.7% Fe, respectively. The flowsheet and material balance diagram showed that the third WHIMS treatment produced concentrate and middlings that assayed 63.5% and 48.6% Fe, respectively. The recoveries for the concentrate and the tailings were 73.1% and 11.5%, respectively.[9]

WHIMS can be used in the following applications:

a. Ferrous metal and ferroalloy ores
   Recovery of concentrates of hematite, limonite, siderite and chromite from their ores.

b. Nonferrous metal ores

Separation of finely embedded wolframite particles from quartz, pyrite from cassiterite in a complex sulphide ore and cassiterite, wolframite from limonite ore. It can also be used to separate wolframite from garnet and in the concentration of iron and tantalum-niobium ores.

c. Rare earth metal ores

Concentration of monazite and phosphorus yttrium ores.

Separation of lithium pyroxene from hornblende, tantalum from niobium, iron ore from microlite, titanium bearing iron ore from rutile, and rutile from garnet.

d. Non-metallic ores

Purifying of industrial raw materials such as quartz, feldspar and kaolin used in producing glass ceramics.

Purifying of high-temperature refractory silicates such as alusite and kyanite.

Elimination of harmful impurities such as iron, hornblende, mica, electrical stone and garnet from ores.

e. Other applications

Treatment of steel mills and power plants wastewater.

14.10   Wet Magnetic Separation with BOXMAG Rapid Equipment

A typical magnetic separator is a rapid disc magnetic separator (25459 Model, 4-3-05.09 Type) shown in Figure 14.4.[10] In the separation process, the slurry is introduced down into the filter at an average velocity of 70 mm/s to separate the magnetic particles in an iron-dominated copper ore. Some nonmagnetic particles are misplaced into the magnetic fraction due to clogging and mechanical trapping of these particles in the matrix.

14.11   Permanent Drum Magnetic Separator

When the ore feed gets to the magnetic drum, the magnetic particles get attracted and stuck to the drum shell. As the drum rotates, the ore particles are carried to pass the stationary magnetic field. The non-magnetic particles then fall freely off from the drum shell while the magnetic particles remain held firmly to the shell until the magnetic field is passed. The principle of operation of the drum magnetic separator is shown in Figure 14.5.[8]

The separation of ore particles by the drum magnetic separator occurs by the pickup principle. Ferro and paramagnetic particles are lifted from the ore charge in motion by the magnet and get pinned to the drum. Field intensity of up to 0.70 T at the magnetic pole surfaces can be obtained in this type of magnetic separator.

FIGURE 14.4   Wet rapid magnetic separator, NMDC, Jos (Ayodele, 2020).

14.12   Low-intensity Magnetic Separator (LIMS)

Metso low-intensity magnetic separator (LIMS) are used for the recovery of pre-concentrates or concentrates of ferromagnetic ores. They are manufactured in three types using a common magnetic drum assembly based on the tank designs, namely concurrent, counter-current and counter-rotation. In LIMS, permanent magnets are used to obtain the highest magnetic attraction force with the highest efficiency. LIMS is used in the concentration of ferromagnetic materials such as magnetite and pyrrhotite as well as in the removal of tramp iron. It is also used to recover magnetite or ferrosilicon in heavy media processing of ores.[1, 11]

Adjustable
deflector regulates
volume of flow with
alternative feed
hopper

Access and
Inspection Opening

Tramp iron pulled
to revolving shell by
fixed magnetic field

Shell revolves
around fixed
magnetic field
Stationary
permanent magnetic
assembly

Magnetic material
held to shell until it
carries past
magnetic field

Adjustable divider

Mounting Channel

Cleaned non-
magnetic material
fall here

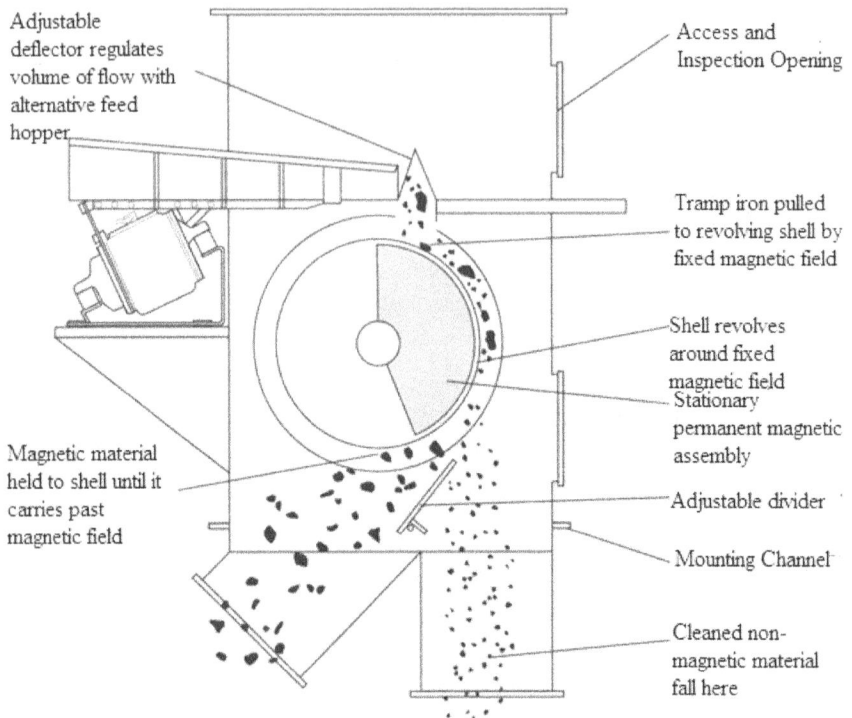

FIGURE 14.5   Drum magnetic separator.

Source: **Courtesy of Eriez (2022b), adapted**

In a concurrent type drum separator, the concentrate obtained is carried along by the drum in the same direction it rotates and passes through a gap where it is compressed and dewatered before exiting the separator. However, in a counter-current type drum separator, the feed flows in the direction opposite to the drum rotation. In a counter-current drum separator, the tailings are made to move in the direction opposite to the one in which the drum rotates and then get received into the tailings chute. The variables affecting the collection of ferromagnetic materials in a wet drum magnetic separator are:

a. The magnetic field strength must be sufficiently high so that ferromagnetic minerals can effectively be collected.

b. Hydraulic capacity-the rate of recovery of ferromagnetic materials is directly proportional to the flow rate of the feed through the separator. The slurry velocity increases as the slurry flow rate increases and this results in the increase of fluid drag force. The latter promotes the detachment of more ferromagnetic particles from the magnetic field acting in opposition.

c. Per cent solids—the ore feed per cent solids have direct effects on how selective the separation is. The slurry becomes more viscous as the percentage of

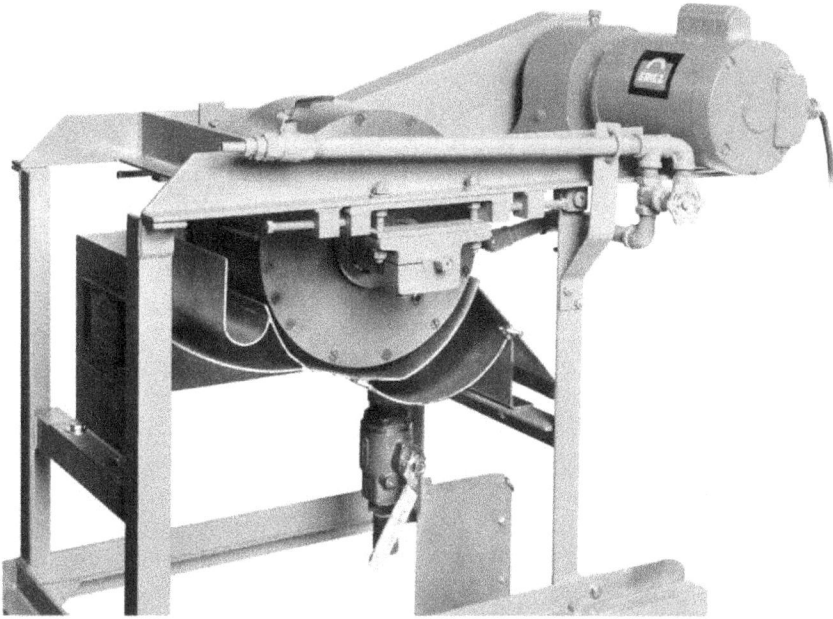

FIGURE 14.6   Eriez wet low-intensity magnetic separator.

Source: Courtesy of Eriez (2022b)

solids increases and this reduces the effects of the fluid drag thus assisting in silica separation.

d. Ferromagnetic content of the ore—the extent of ferromagnetic material removal in a wet drum magnetic separator is restricted by the diameter of the drum, the applied magnetic field strength and peripheral speed. This phenomenon is called "magnetic loading". Any attempt to exceed the limits dictated by magnetic loading will lead to increased magnetite losses.

Figure 14.6 shows Eriez wet low-intensity magnetic separator.[8]

A typical LIMS circuit may consist of rougher, scavenger and cleaner units. It can receive a spiral concentrate for further treatment. Its capacity depends on the volumetric flow rate of the ore it can take.

14.13   Laboratory-scale Magnetic Concentration of Ores

Setlhabi et al. (2019)[12] reported the use of an Eriez dry high-intensity magnetic separator, size 380 × 160 mm shown in Figure 14.6 to separate the silica gangue in a chromite mineral sample that assayed about 35%. About 500 g of each size fraction having 10% moisture content was fed into the separator for each test at 15 rpm using a vibration feeder frequency of 8 Hz. The paramagnetic chromite mineral concentrate obtained was dried in the oven at 80°C, weighed and packaged for XRF analysis.

Ayeni et al. (2012)[13] reported the recovery of columbite concentrate from Nigeria Jos Rayfield tailings dump secondary resources. About 55 kg of the columbite tailings obtained by grab sampling was screened 100% passing a 1 mm screen. 50 kg of the sample <1 mm was fed into a dry high-intensity magnetic separator (Rapid) Model 4-3-15 OG, at a feed rate of 250 kg/h and at varying applied currents of 0.2, 0.5 and 1.0 A at the discs for magnetite, hematite columbite recovery, respectively. The sample was received into the hopper and the shutter was slightly opened for slow and even spread of the ore feed on the magnetic belt that conveyed the feed through the three discs of the rapid separator. The ferromagnetic magnetite, paramagnetic hematite interlocked with columbite and the paramagnetic columbite were separated at the first, second and third disc, respectively.

For maximum recovery, the hematite middlings concentrate that was rod milled to obtain 100% passing 355 μm was further reprocessed using a pneumatic (air) floating table, Kipp Kelly model MY to recover rougher concentrate, middlings, and tailings. The middlings was recycled to the Kipp table. The pre-concentrate from the rapid was further processed using an air-floating machine operated at a tilted angle of 60°, speed of 2.75 stroke/s, motor speed of 1425 rpm and feed rate of 250 kg/h. The products of the air floating were the final concentrate of columbite, middlings and tailings. The air flotation process was further repeated for the middlings and tailings while the new output tailings were added to the tailings obtained from the rougher process to be the final tailings of the processing. Samples were taken from products of each processing unit for analysis using an ED-XRFS machine (Energy Dispersion-X Ray Fluorescence spectrometry) Minipal 4 Model 20. Figure 14.7 shows Eriez dry high-intensity magnetic separator.[14]

14.14   Pilot-scale Magnetic Concentration of Ores

Ola et al. (2009)[15] processed the Itakpe sinter-grade concentrate through a Boxmag-Rapid wet low-intensity magnetic separator at a rate of 250 kg/h capacity to take away the ferro-magnetic magnetite component so that a residue comprising a combination of the mainly diamagnetic silica gangue and para-magnetic hematite can be obtained. The residue was then delivered to a Humboldt-Wedag wet high-intensity magnetic separator operating at 1.25 T to separate the para-magnetic hematite from the tailings. The concentrates of the wet low-intensity magnetic separation (LIMS) and the wet high-intensity magnetic separation (WHIMS) were then combined to form the magnetic super-concentrate or Midrex Concentrate. The mass balance for the process is shown in Figure 14.8.[15]

14.15   Magnetizing Roasting Experiment

A hematite ore can be roasted to become ferromagnetic. About 650 g of the sample is put in a silica dish $145 \times 85 \times 20$ mm to occupy three-quarters of the volume. The silica dish and the content are then placed in a cold furnace set at 350°C and then switch on. The electric furnace is opened 25 mm wide during the experiment

FIGURE 14.7   Eriez dry high-intensity magnetic separator, TUT, South Africa.

to allow air in. The sample is turned in every 15 min. The temperature is raised by 100°C every hour in 4 steps of 15 min till 750°C. The sample is left in the furnace for 2 more hours at 750°C while still stirring every 15 min.[16]

14.16   Electrostatic Separation

The electrostatics equipment consists of an earthed mild steel roll also called a rotor. It is equipped with an electrode assembly having a rectified DC voltage of up to 50 kV. The electrode assembly ionizes the surrounding air to produce electron charges. The ore particles are first delivered into the hopper where they are heated to full dryness.

The electron charges produced are used to spray the ore particles to be delivered unto the roll. The non-conducting or poor conducting mineral particles receive the electric charges and largely retain them and the large surface charges get them attracted to and pinned to the rotor surface. On the other hand, conducting mineral particles or those with comparatively higher conductivity receive the charges and

Itakpe Sinter Concentrate 77% < 500 μm

The Key Box

| 100 | 63.70 | 100 |

LIMS

| | %Fe tot | |

wt% Feed recovery   wt% Fe recovery

LIMS Superconcentrate I

| 13.20 | 67.59 | 14.00 |

Paramagnetic Hematite

| 86.80 | 62.83 | 85.61 |

WHIMS

Superconcentrate II

| 78.73 | 69.70 | 84.91 |

Middlings

| 0.01 | 60.60 | 0.01 |

Tailings

| 8.06 | 5.50 | 0.70 |

Itakpe Superconcentrate

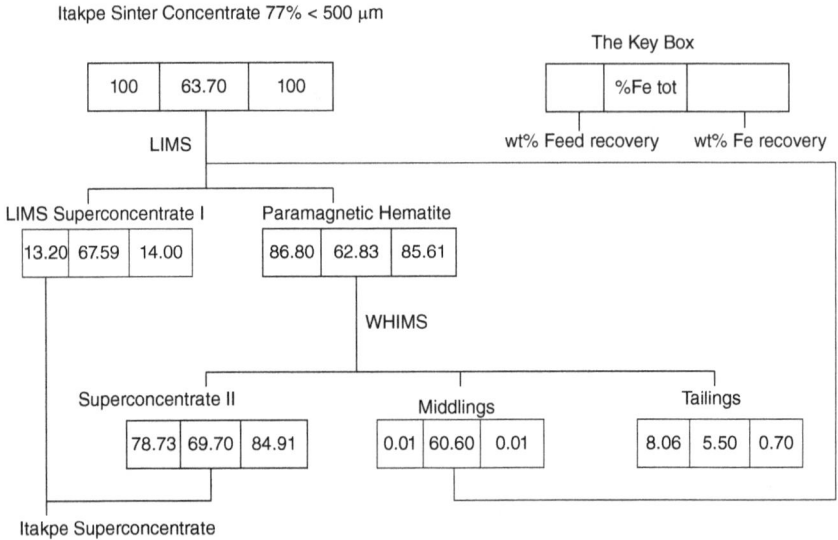

FIGURE 14.8 Mass balance for the production of Itakpe midrex concentrate by LIMS-WHIMS magnetic separation.

rapidly dissipate them unto the rotor and they continued approximately in a path they would have taken without the charge.

High-tension separators take in feeds containing particles with sizes between 60 and 500 μm diameter. Particle size has effects on the separation behaviour since the surface charges' influence on a coarse non-conducting or poor conducting grain is lower in relation to its mass in comparison to the effect on a fine grain. Thus, a coarse non-conducting grain is more readily thrown from the roll surface instead of being pinned to it and is misplaced into the conducting fraction. Similarly, the finer conducting particles are more affected by the surface charges than the coarse ones and are thus more prone to being misplaced into the non-conducting fraction.

Figure 14.9 shows the principals involved in electrostatic separation.[17]

In the plate-type electrostatic separators, the dry feed is passed over a charged anode that attracts electrons from a vibratory feeder to receive electron charges. The charged particles then move onto a sloping, grounded plate and the conducting particles release their charges unto the plate and get separated from the non-conducting particles which are inductively attracted to the overhanging static electrode assembly. They then follow a path different from those of the non-conducting particles. It is widely used to process mineral sand.

Wills and Napier-Munn (2006)[1] reported examples of the non-conducting particles that are pinned to the rotor and conducting particles that are thrown as follows:

Examples of non-conducting minerals that are pinned to the roll (rotor):

a. Apatite, barite, calcite, corundum, garnet, gypsum, kyanite, monazite, scheelite, sillimonite, spinel, tourmaline, zircon and quartz
b. Conducting minerals that are thrown from the roll:

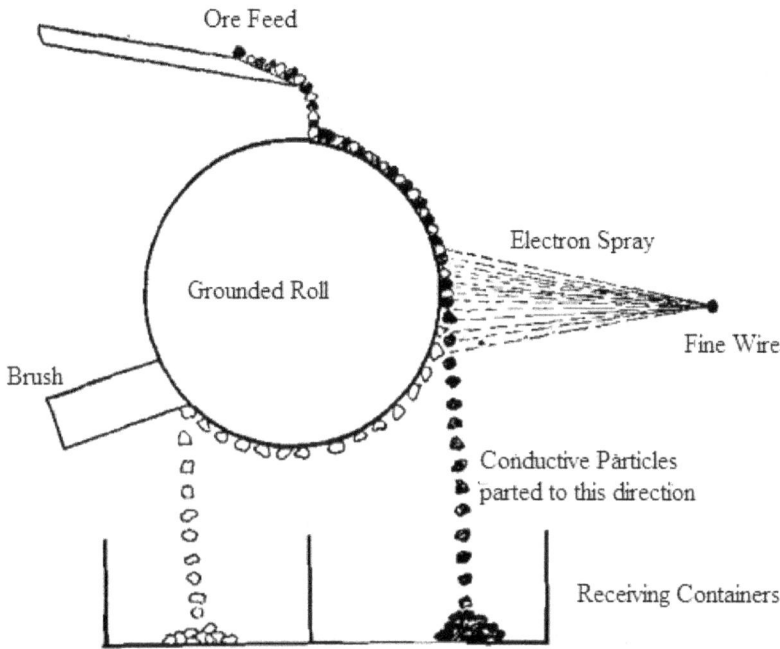

FIGURE 14.9   Electrostatic separation method.

Source: **Courtesy of 911 Metallurgist (2022c)**

Cassiterite, chromites, diamond, fluorspar, galena, gold, hematite, ilmenite, limonite, magnetite, pyrite, rutile, sphalerites, stibnite tantalite, wolframite

References

1. Wills, B.A., and Napier-Munn, T. (2006): *Mineral Processing Technology-An Introduction to the Practical Aspects of Ore Treatment and Mineral Recovery*, 7th Edition. Amsterdam: Elsevier Science and Technology Books, p. 450.
2. TAMU- Magnetic Susceptibility-Ocean Drilling Program (https:www-odp.tamu.edu/put, Accessed 23rd July, 2020).
3. Natalia Petrovskaya, Classification of Minerals on the Magnetic Properties. *LinkedIn* (https://www.linkedin.com/in/ph-d-natalia-petrovskaya).
4. Wikimedia (2022a): *Magnetic Separation* (https://en.wikipedia.org/wiki/Magnetic_ separation, Accessed 5th May, 2022).
5. 911 Metallurgist (2022a): *Magnetic Separator/Filter-Industrial Filtration Systems* (https://th.bing.com/th/id/OIP.U7XQWb-q4cgNRe1pfpIKugHaDM?pid=ImgDet &rs=1, Accessed 30th June, 2022).
6. 911 Metallurgist (2022b): *Wet Drum Magnetic Separator* (www.911metallurgist.com/ equipment/wet-magnetic-drum-separator/, Accessed 5th May, 2022).
7. Eriez (2020a): (https://wpeprocessequipment.com.au/equipment/eriez-whims/, Accessed 25th July, 2020).
8. Eriez (2022b): (www.eriezlabequipment.com/lab-equipment/magnetic-separators/wet-low-intensity-magnetic-separators-wlims/, Accessed 25th July, 2020).

9. 911 Metallurgist (2022c): *Magnetic Separators* (www.911metallurgist.com/equipment/magnetic-separators/, Accessed 30th June, 2022).

10. Ayodele, T.J. (2020): *Optimization of Hydro Extraction of Copper from a Hematite Dominated Copper Ore*. PhD Qualifying Examination Report, Department of Materials Science and Engineering, Obafemi Awolowo University, Ile Ife, Nigeria Discourse (https://ask.learncbse.in/t/draw-the-diagram-showing-i-froth-floatation-ii-magnetic-separation/10062/1, Accessed 13th July, 2020).

11. Eriez(2020c): (www.eriez.com/NA/EN/Products/Magnetic-Separation/Permanent-Magnets/Drum-Separators.htm, Accessed 25th July, 2020).

12. Setlhabi, B.I., Popoola, A.P.I., Tshabalala, L., and Adeleke, A. (2019): Evaluation of Advanced Gravity and Magnetic Concentration of a PGM Tailings Waste for Chromite Recovery. *Iranian Journal Chemistry and Chemical Engineering*, 38(2), 61–71.

13. Ayeni, F.A., Madugu, I.A., Sukop, P., Ibitoye, S.A., Adeleke, A.A., and Abdulwahab, M. (2012): Secondary Recovery of Columbite from Tailing Dump in Nigerian Jos Mines Field. *Journal of Minerals & Materials Characterization & Engineering*, 11(6), 587–595.

14. Eriez (2020d): (www.eriezlabequipment.com/lab-equipment/magnetic-separators/dry-high-intensity-magnetic-separators-dhims/, Accessed 26th July, 2020).

15. Ola, S.A., Adeleke, A.O., Usman, G.A., Odunaike, A.A., Kollere, A.M., and Ajiboye, P.O.(2009): Study on Magnetic Concentration of Itakpe Sinter Concentrate to a Midrex Grade Concentrate. *Maejo International Journal of Science and Technology*, 3(3), 400–407.

16. Jianwen, Y., Yuexin, H., Yanjun, L., and Peng, G. (2017): Beneficiation of Iron Fines by Magnetic Roasting and Magnetic Separation. *International Journal of Mineral Processing*, 168, 102–108.

17. 911 Metallurgist (2022e): *Electrostatic Separation Method* (www.911metallurgist.com/wp-content/uploads/2017/03/Electrostatic-Charges-Delivered-by-Electric-Spray.png, Accessed 2nd April, 2022).

15 Ore Sorting

15.1 Introduction

In ore sorting, each ore particle is appraised for acceptance or rejection. Ore sorting is primarily designed to separate ore minerals from gangue or low-grade material in the early part of the separation process. Manual sorting was the common practice until the 1940s when the constraints of technology that made the earlier envisaged automated sorting impossible were removed. The benefits of ore sorting include:

a. The rejection of low-grade or barren rock particles after crushing will make it possible to set up processing plants with smaller comminution and separation facilities and this will lead to a reduction in capital and operating costs.
b. The available plant treatment capacity determines the mining rates possible. Ore sorting will reduce the tons of materials to be treated and increase the capacity to take more runs-of-mine without a major expansion of the existing facilities.
c. Ore sorting can help in upgrading a deposit with a grade previously considered too low to be processed for economic processing.
d. Ore sorting can assist in the qualitative grading of products based on a given physical characteristic the sorter uses. For instance, if a sorter sorts based on color, it can produce several products of different sale values and thus enhance sales income.

15.2 The Principles of Electronic Sorting

The principles that underlie ore sorting can be summarized as follows:

a. Sorting is mainly applicable to ores that are liberated at coarse sizes greater than 5–10 mm so that the gangue waste can be thrown away with not much valuable minerals loss.
b. Sorting is a pre-concentration technique used to reduce the tonnage of materials to be treated in the downstream concentration processes.
c. Properties of rocks such as reflectance, the color reflected in visible light, magnetism, conductivity, microwave attenuation and other properties are used. For instance, color in visible light is used in sorting magnesite, dolomite, base metals and gold ores while conductivity is used for sulphides and natural gamma radiation for uranium ores.

15.3 Mechanism of Operation

An electronic sorter inspects the ore charge particles to determine the value of a rock property for each particle. It then ejects a particle, valuable or gangue, based on some criterion of the property selected for which a distinct difference occurs between the

DOI: 10.1201/9781003323433-15

value and the gangue particles. To avoid signal blurring, the sized particles must be washed free of dirt and should be charged in a single layer to ensure the exposure of each particle to the machine.

15.4   Types of Ore Sorters

There are four major types of mechanized ore sorters in relation to the physical property of the rock particles to be used in the detection operation. These are electronic, photometric, radiometric and conductivity-resistivity sorters. The ore sorting process involves singulation, detection and ejection. Singulation refers to the flow of the ore particles or fragments into the detector for inspection. The detector then senses the presence or absence of the desired fragment and ejection effect the physical separation of the detected fragment from the remaining fragments in the ore charge. Photometric sorters use an optical property such as color or reflectivity for sorting, while a radiometric sorter detects the presence or absence of gamma rays to sort. The conductivity-resistivity sorter determines the conductivity or resistivity of a fragment and compares the results to a standard to accept or reject the rock fragment.

As a practical example, in the electronic sorter applied to copper ores, ore fragments are introduced into the electric field of a tuned coil oscillating at high frequency. If a fragment contains enough copper, the frequency of the oscillation becomes altered and this will trigger its ejection into the pre-concentrate. For the electronic sorting to be successful, the run-of-mine rocks must have a high degree of heterogeneity and must be of the appropriate size.

The automatic Photometric Sorter is a mechanized form of hand sorting and the sorter has a laser source and a sensitive photomultiplier. The sorter releases laser light on the ore particles' surfaces and the particles' reflected light is taken in by the photomultiplier to produce signals which change based on the intensity of the reflected light. An electronic system analyses the photomultiplier signals to produce control signals through which appropriate valves of an air-blast rejection system are actuated to eject certain particles based on the analyses carried out.[1-4]

15.5   The Model 16 Photometric Sorter

The Model 16 is designed to sort value ore mineral particles that exhibit reliable and consistent optical difference from their associated gangue minerals and it can handle particles with sizes ranging from 10 to 150 mm. It works by determining the light reflectance emanating from each particle and can therefore be used for many minerals.

For efficient sorting, the feed material has to be properly prepared by removing fine-sized constituents in order to improve light reflectance properties and screening to ensure that the ratio of the largest to the smallest particle is about 3:1. The rock feed system is required to efficiently present each particle for scanning and separation. The system is of high capacity as it can perform 2000 scans per second.

The separation system is equipped with high-speed solenoid air valves that are independent of each other. The scanner sends signals to the air valve to indicate the selected and rejected rocks. The selected rocks are deflected from their free-flight

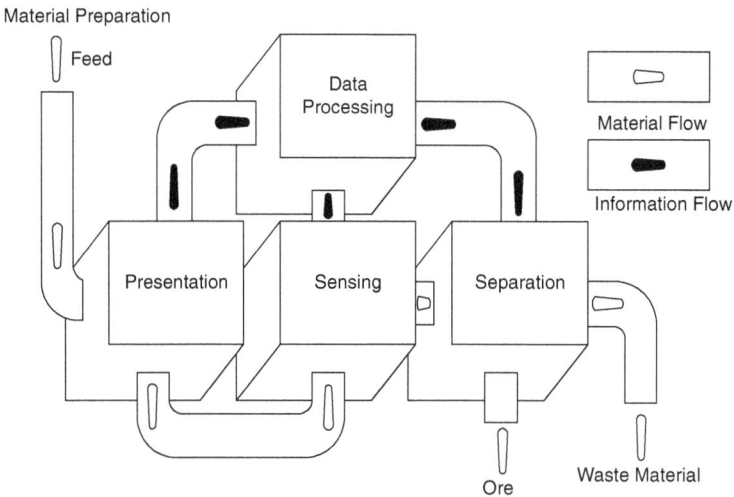

FIGURE 15.1 Photometric sorter sub-system.

Source: **Courtesy of 911 Metallurgist (2022a)**

trajectory path by high-pressure air blasts into a stream and the rejected rocks are diverted into another stream both separated by a splitter plate. Finally, the rocks on the two distinct paths are carried by two chutes to ore and waste conveying belts. Model 16 Photometric Sorters have been used to process various ores such as gold reef, wolframite in quartz, spodumene, copper/silver in quartz, phosphate, limestone and wollastonite. Figure 15.1 shows the sorter sub-system of the Model 15 sorter.[2, 5]

15.6   The Belt Sorter

The Belt Sorter is an electronic sorter. It is a conveyor belt placed over a tuned coil oscillating at high frequency upon which pieces of ore pass. If, for instance, a rock contains sufficiently high native copper content, an alteration of the frequency of the oscillation occurs. An electronic circuit detects the change in the frequency and opens an air valve positioned to deflect the copper-bearing piece from its normal trajectory at the end of the belt and onto a concentrate conveyor. Figure 15.2.[5] shows the schematic diagram of the operation of a belt sorter.

15.7   Industrial Ore Sorting

Kgaka et al. (2015)[6] reviewed the operation of a South African gold mining company that sorts a dump of waste rocks with gold content that ranges between 0.2 and 0.9 g/t using the Commodas optical sorter. The sorter is used to reduce the gangue content mixed in the waste rock to the barest minimum prior to further treatment. The results obtained showed that the Commodas sorting machine was not operating at its 70% gold recovery set target consistently. The studies revealed that the highest and second

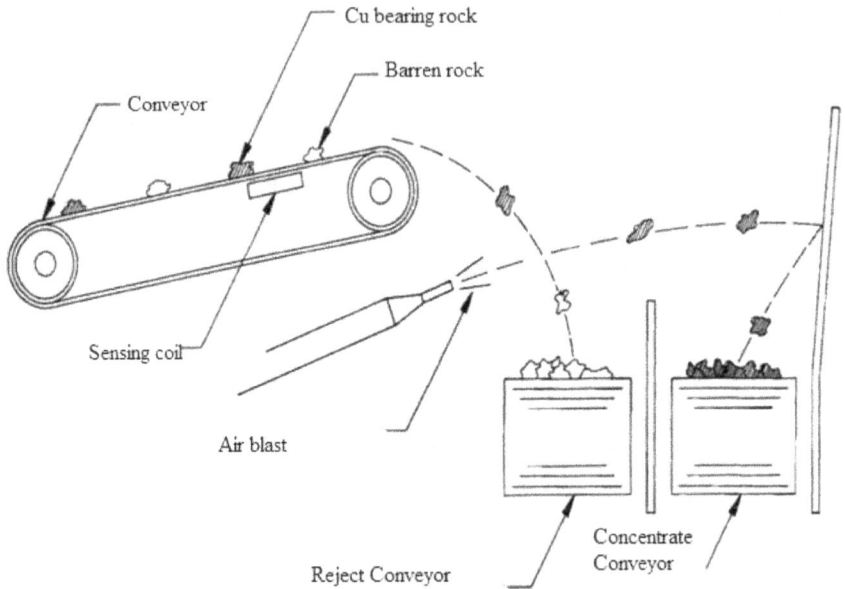

FIGURE 15.2   Schematic diagram of belt sorter operation.

Source: **Courtesy of 911 Metallurgist, 2022(b).**

highest gold recovery per cents of 71.2% and 61.6% were obtained when the average rock feed rates were 24.8 and 26 tph, while very low recoveries of 2.4% and 3.1% occurred at the feed rates of 46 and 29.5 tph, respectively. It was therefore proposed that the waste rock feed rate into the Commodas optical sorter should be maintained in the range of 24.8–26 tph and the process parameters be further studied and adjusted so that gold recovery per cents will be fairly proportional to the feed grade and mass pull into the concentrate and inversely proportional to the waste stream grade.

References

1. Wills, B.A., and Napier-Munn, T.J. (2006): *Mineral Processing Technology*, 7th Edition. Amsterdam: Elsevier Science and Technology Books.
2. 911 Metallurgist (2022a): *Sorting of Copper Ore* (www.911metallurgist.com/sorting-copper-ore/#:~:text=The%20four%20major%20types%20of,of%20the%20rock%20for%20separation, Accessed 27th July, 2020).
3. Avant Mining (www.avant-mining.com/our-business/innovation-and-technology/).
4. AUSIMMBulletin(2020):(www.ausimmbulletin.com/feature/sensor-based-ore-sorting-maximise-profit-gold-operation/).
5. 911 Metallurgist (2022b): *Photometric Sorter Sub System* (https://z4y6y3m2.rocketcdn.me/wp-content/uploads/2018/10/Photometric-Ore-Sorting-Sub-System.jpg, Accessed 22nd April, 2022).
6. Kgaka, M., Adeleke, A.A., Popoola, A.P.I., and Olubambi, P.A. (2015): An Evaluation of the Efficiency of Commodas Optical Sorting of Gold Waste Rock Dumps. *Mineral Processing & Extractive Metallurgy Review*, 36, 123–128.

16 Dewatering and Tailings Disposal

16.1 Introduction

Most mineral processing operations of ores are conducted wet with the solid ore particles carried as slurries, a mixture of water and solids. The final concentrated solid particles, therefore, need to be separated from the carrier water content.[1] Methods of dewatering may be categorized into three types—filtration sedimentation and thermal drying.

Dewatering involves the separation of water from solids for one of the following reasons:

a. To prepare the slurry for subsequent processing that requires higher solids content
b. To obtain a reduced slurry volume for downstream processing
c. To obtain a concentrate product with a target per cent moisture

In clarification, a waste slurry is treated to recover water of the correct quality for reuse in the processing plant. For instance, thickeners are used to clarify wastewater for reuse and for the thickening of slurry.

16.2 Filtration

In filtration, suspended solid particles in a liquid are separated by causing the carrier liquid to pass through a filter to produce a filtrate, that is, the liquid that passed through the filter and a residue of solid particles retained on the filter. There are three types of filtration techniques-mechanical, biological and chemical filtration. Filtration involves the use of filters to maximize the recovery of valuable particles with sizes below 325 mesh (44 μm). Pressure filters are used to obtain high particle recovery and acceptable moisture contents. Figure 16.1 shows the Eriez filter press. They are used to dewater slurries and to separate solids from liquids.[2]

The required area and number of filters to be used depend on the amount of filtrate to be separated from the product and the desired moisture content.

16.3 Sedimentation

In wastewater treatment, sedimentation is described as a physical process that uses gravity to remove suspended solids from water. It is also described as the tendency for particles suspended in a fluid to settle out of the fluid under gravity and attain rest against a barrier as sediment in mineral processing or sludge in wastewater

DOI: 10.1201/9781003323433-16

FIGURE 16.1   Eriez MACSALAB filter press.

Source: Courtesy of Eriez (2022a)

treatment. The efficiency of sedimentation reaches its peak when there is a large difference in the density of the carrier liquid and the solid particles in suspension and this condition is obtained when water is the carrier liquid.

On the other hand, the leach liquor in the hydrometallurgical processing of low-grade ores is usually of high densities which are close to those of the solids in suspension. In such cases, filtration may be required instead of sedimentation. When the solids in suspension settle under gravity in a tank, usually cylindrical in shape, to form a thick pulp, the sedimentation process is called thickening. Thickening is also accomplished by decanting as a preliminary process.

16.4   Thermal Drying

Thermal drying involves the de-hydration of moist products by evaporating their moisture using heat. The most common type of dryer in mineral processing are the convection dryers such as drum, conveyor and fluidized bed dryers. They use the heat from hot combustion gases to extract moisture from a slurry stream.[3]

16.5   The Principles and Operation of a Thickener

A thickener is a cylindrical equipment that separates water from solids in a slurry by gravity sedimentation process. Thickening can be defined as a method of continuous de-watering of a dilute slurry where a continuous discharge of a thickened pulp of uniform density with particles size lower than 0.5 mm takes place concurrently with the overflow of a clarified solution. It is equipped with rakes or scraper blades which rotate slowly over the bottom of the tank and slope down towards the centre. It moves the particles settled on the bottom to a central opening or discharge. Thickeners are used for the following purposes:

a. Thickening of concentrates
b. Thickening before slurry agitation
c. Thickening before filtration in the countercurrent washing of cyanide slimes
d. More thickening prior to froth flotation
e. For the de-watering of tailings to recover the water content for re-use in the mill

Both thickeners and clarifiers are used to settle solids in slurries resulting in the separation of solids and liquids from each other. Thickeners are used to concentrate solids, while clarifiers are used to purify liquids. In continuous thickeners, a slurry is fed into the centre of the tank through the feed well which is placed up to a depth of 1 m below the suspension. The clarified liquid overflows a trough, while the settled solids are removed as thickened pulp from an opening at the centre. The thickener is equipped with some radial arms each of which is a series of blades with a shape designed to rake the settled solids towards the central outlet. The basin is constructed of steel plate or concrete.[4]

In practice, de-watering in mineral processing is usually a combination of the methods mentioned. The major portion of the water is first taken away by sedimentation and decantation or thickening to obtain a pulp thickened to contain 35–45% water by weight and this implies up to 80% of the water separated. Afterwards, the pulp is filtered to obtain a moist cake with moisture content ranging from 10% to 20%. The thermal drying of the latter will produce a final product of about 10% moisture or 90% solids.

16.6   Coagulation and Flocculation

Flocculation is a two-step particle aggregation process in which a large number of fine particles form a small number of large flocs. The first step is called coagulation and the second step is flocculation. Coagulation is applied prior to sedimentation or filtration to enhance the removal of fine particles. It is used to render surface charges neutral to form a gelatinous mass to trap particles and thus form a mass large enough to settle or trap in a filter. Fine-sized particles are usually hindered in the quest to aggregate and settle because they carry negative surface charges.[5]

When coagulation chemicals carrying positive charges are added to the slurry, the molecules adsorb on the fine particles to cause charge balancing. The introduction of the charges with opposite polarity stimulates the formation of stable and well-suspended sub-micron flocs from ore particles that stick together. Mixing the slurry rapidly is necessary to disperse the coagulant chemical molecules and promote particle collisions and sub-micron flocs formation.

Flocculation refers to the gentle mixing to stimulate the masses formed during coagulation to agglomerate to larger masses that can settle or get trapped in a filter. The flocculation step requires not only gentle mixing but also the use of polymeric flocculant of high molecular weight. The adsorption of flocculant on the flocs of the sub-micron scale causes the gaps between flocs to be bridged. The coming of flocs together produces a range of gap for Van der Waals attraction forces to become effective in reducing the energy barrier for flocculation and this leads to loosely packed flocs forming.

This is followed by aggregation binding and strengthening of the flocs until the formation of visible macroflocs. When the right weight, size and strength are obtained for the macroflocs, sedimentation occurs. In mineral processing and hydrometallurgy, selective precipitation of metals from solutions is also achieved by flocculation and sedimentation. Salts of aluminium such as its chloride, sulphate and aluminate and iron-based salts like ferrous sulphate, ferric sulphate and ferric chloride are coagulants used for wastewater treatment. The diagrams presented on the mechanisms of coagulation and flocculation in Mettler Toledo (2020)[5] elucidate the two processes very clearly.

16.7 Tailings Disposal

Tailings disposal remains a challenge in the mineral processing industries because of the following reasons:

a. The environmental impacts and risks associated with tailings that necessitate stringent legislation on tailings disposal
b. The increasing need to mine and process low-grade ores has increased water consumption per unit of production and this necessitates recovery of tailings water for re-use
   Tailing treatments led to the overflow discharge of water fouled with solid particles, metals of high density, reagents for the milling operations, sulphur-based compounds and others. Tailings waste residue disposal also defaces the landscape and must meet extant legislations to protect the environment. The old methods of tailings disposal include:
c. Discharge into streams and rivers
d. Dumping of de-watered coarse tailings unto a land

In view of the ecological; damage caused by the old methods, the following satisfactory methods are now used for conventional tailings disposal:

a. Re-processing the finely ground ore for additional recovery of values
b. Convert tailings to useful products such as railway ballast and aggregate
c. Return of coarse tailings to fill up mined-out areas in underground mining as backfill
d. De-sliming of the tailings and surface disposal of the slimes
e. The large tailings resulting from open pit mining are usually disposed into dams

In conventional tailings disposal, the mill slurry with 20–45% solids or the shallow thickener with 30–55% solids by weight is disposed of as mentioned without further treatment. However, the need to minimize water consumption has made mineral processing plants to develop alternative tailings disposal (ATD) methods. The ATD includes:

a. Filtered Tailings Disposal
   This involves removing water by vacuum or pressure methods. Tailing slurries are de-watered by vacuum using drum, disc or belt filters or alternatively by pressure with filter presses or belt press filters to produce dry cakes with over 85% solids by weight.
b. Paste Tailings Disposal
   The tailings slurry is de-watered in specialized paste thickener or ultra-high-density thickener to a pulp density of 70–85% solids by weight suitable for pumping and is transported away with a centrifugal pump.
c. Thickened Tailings Disposal

In this method, thickened tailings with 50–70% solids by weight are produced in a deep cone or high-density thickener to be disposed of by centrifugal pump.

16.8   Laboratory Testing for Thickener Design

The capacity of a metallurgical sedimentation device such as a thickener depends on the following factors:

a. slurry feed rate and its associated settling rate
b. % solids or slurry feed dilution
c. suspended solids size and shape
d. the solid and liquid phases' specific gravity differential
e. electrolytes and/or flocculants presence
f. viscosity of the pulp
g. temperature of the slurry

Experiments are typically carried out in graduated cylinders to show different stages of particle or slime settlement as a series of sedimentation phases that provide mudline height with respect to time. The slurry is first thoroughly mixed and allowed to settle over time and produces four zones—A, B, C and D. Zone A represents the region of a clear solution formed, while zone B refers to the region having the original pulp minus any coarse particle that has settled. There may also be a transition zone labelled as C and finally the compression zone D. Zones B and C are of free settling origin, unlike zone D. The relative thickness of each zone varies with the pulp type. The critical point is the point just after zone B and/or C just disappear or when zone A and zone D just get in contact and the slurry attains its highest density. A plot of midline height and time is then constructed.

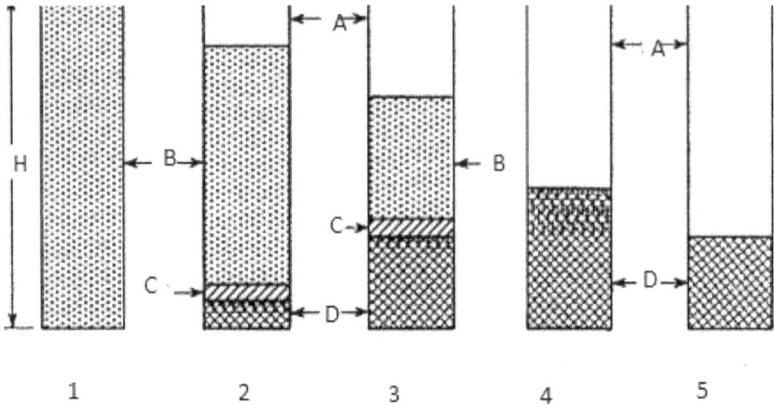

FIGURE 16.2   Progress in batch sedimentation with settling zones.

Source: McCabe et al. (1993) (Courtesy of Elsevier Book Publisher, 2022)

Based on experimental observations, the following zones are identified in a thickener as shown in Figure 16.2[6]:

Zone A: Clear supernatant liquid
Zone B: Free settling zone
Zone C: Transition zone
Zone D: Compression zone

Coe and Clavenger reported that the capacity of a continuous thickener depends on the % solids in the thickener different from that of the feed pulp. They derived the unit area of solid sediment obtainable A as:

$$A = \frac{1.33(F-D)}{Rr} \qquad (16.1)$$

where
   A = unit area requirement for sedimentation in square ft2/ton/day (square feet per ton of dry solids per day)
   F = feed dilution
   D = underflow dilution
   R = rate of subsidence
   r = specific gravity of the liquid

From a tabulation of the unit area and the corresponding feed dilution, the largest unit area is selected as the basic design criterion

16.9   Industrial Thickener

An industrial thickener consists of a cylindrical tank with a 15–30 m diameter, a depth of 2–3 m and a tapered conical bottom. The slurry treated in the plant is fed into the thickener at the concealed centre at a depth of about 1 m below the liquid surface avoiding the content disturbance as much as possible. The thickener has a slowly rotating rake carrying scrapers at the bottom of the tank. The thickener rotates at the drum and conical section with a speed of one rotation at 69 and 76 s, respectively. The rake agitates the slurry to reduce the apparent viscosity of the suspensions and effects the compression of the thickened material and moves it towards the outlet. The thickened slurry, called sludge, is continuously pumped to the tailings dam or discard dump while the clarified water runs into a launder. Flocculants are added to aid fine particle agglomeration and their fast settling. The concentrate slurry is thickened to a specified pulp density before discharge for drying.

The clarifier acts as a final buffer to remove fine solids from the thickener's overflow streams. The overflow water from the clarifier is re-used in the concentrator plant, while the underflow is taken to the spillage tank. Reuse of water helps in saving costs. Figures 16.3 and 16.4 show well-labelled images of the thickener and inner parts of a thickener, respectively. Figure 16.5 presents a well-labelled image of a clarifier.

In a thickener, a velocity break box receives the high-velocity slurry before delivering it into the thickener feed pipe. In the feed pipe, a flocculant is added to the slurry to produce larger agglomerates that settle to the thickener's bottom in a more

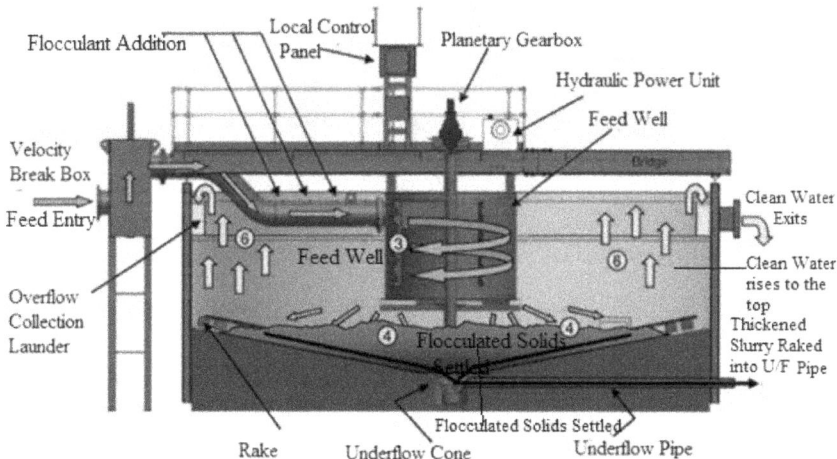

FIGURE 16.3   A labelled thickener.

Source: Courtesy of 911 Metallurgist (2022a)

FIGURE 16.4   The inner part of a thickener.

Source: **Courtesy of 911 Metallurgist (2022a)**

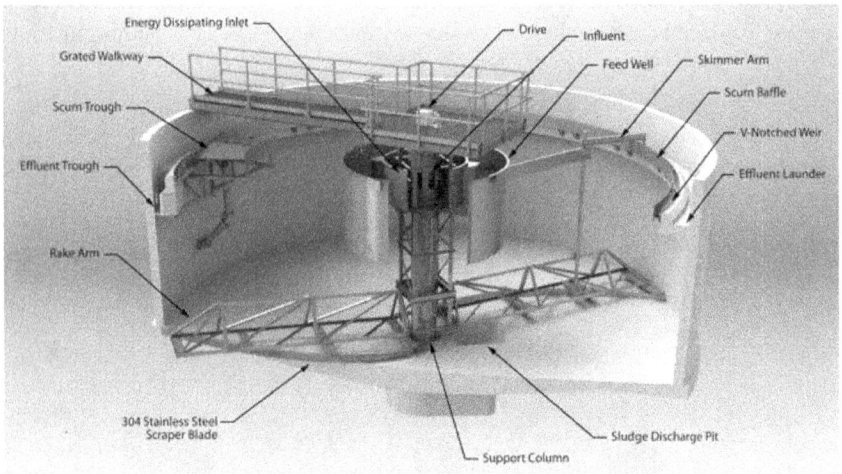

FIGURE 16.5   A clarifier.

Source: **Courtesy of 911 Metallurgist (2022b)**

rapid manner and get compacted to form a slurry of high density in the compression zone. The air entrapped in the flocculated slurry is prevented from entering the main settling zone as it gets dissipated in the oversized feed well. The rake drives the thickened pulp with 35–50% solids to the underflow cone where it is pumped into the underflow pipe. Clean water rises to the thickener's top and overflows into a launder from where it goes into storage or pond for re-use. Figures 16.3, 16.4 and 16.5 show details about the thickener and clarifier.[7, 8]

Thickeners and clarifiers look alike physically but the former is used to concentrate solids and thus produce an underflow with a high percentage of solids, while the latter is applied in purifying liquids to give an overflow that is clearer of suspended solids than for a thickener's overflow. Solid minerals and quarry aggregates are produced by thickeners, while clarifiers are used to extract ore fines and provide water for reuse.

16.10   Talmadge and Fitch Method

In the curve shown in Figure 16.6[9], the slurry height in the test cylinder (H) is plotted against time (t) in seconds. The critical settling point is the point at which the settling curve flattens out and the point tu corresponds to Hu, the mudline height at the critical settling point of the slurry. To find the critical slurry concentration in terms of % solids that control the slurry handling capacity, tangents are drawn to the initial and final legs of the settling curve as shown. The angle of intersection of the tangents is bisected and the bisector is extended to the settling curve to get Cc. The time at which the slurry concentration is Cu is found by drawing a tangent through Cc. From Hu, a horizontal line is drawn to touch the tangent through Cc. A vertical line is then extended from the intersection to the x-axis to obtain tu. Hu corresponds to an underflow concentration of Xu.

16.11   Worked Example

A slurry of 25% solids is to be de-watered in a thickener at a rate of 80 tph to produce a product with 10% moisture. A settling test was conducted on the slurry and the results obtained revealed that the mudline height of 75 mm is the critical point of

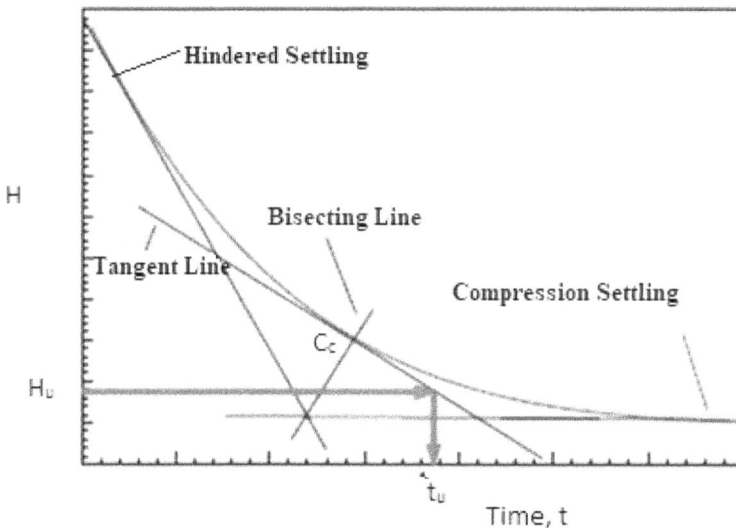

FIGURE 16.6   Slurry height variation with time (Romain, 2012).

the settling curve and occurs after 260 s. The height of the slurry in the test cylinder initially was 290 mm. The density of the suspended solids is 2700 kg/m³ and that of the carrier water is 1000 kg/m³. Determine:

a. The slurry density D
b. The specific gravity of ores in the underflow
c. The thickener area required for the slurry

Solution

$$\text{But } x = \frac{100s(D-w)}{D(s-w)}$$

$$x = \frac{100 \times 2700(D-1000)}{D \times (2700-1000)} = \frac{100 \times 2.7(D-1000)}{D \times (2.7-1.0)}$$

$$25 = \frac{270D - 270000}{2.7D - 2.7}$$

$$67.5D - 67.5 = 270D - 270{,}000$$
$$202.5D = 2669932.5$$
$$D = 1333 \text{ kg/m}^3$$

Therefore specific gravity for the solids $(R_f) = 1.333$ and that of water $(Rw) = 1.0$

$$\text{But } H_u = \frac{H_F(R_F - 1)}{(R_u - 1)}$$

where:

H_u = the mudline height at the critical settling point of settling curve

= 75 mm = 0.075 m

H_F = initial height of the slurry in the cylinder = 290 mm = 0.29 m

R_F = the specific gravity of the slurry in the cylinder = 1.333

R_u = specific gravity of the slurry in the underflow

Therefore:

$$0.075 = \frac{0.290 \times (1{,}333 - 1)}{(R_u - 1)}$$

$$0.075 R_u - 0.075 = 0.09657$$

$$0.075 R_u = 0.17157$$

$$R_u = \frac{0.17157}{0.075} = 2.2876$$

$$x_{u/f} = \frac{100 \times 2700\left(2287.6 - 1000\right)}{2287.6 \times \left(2700 - 1000\right)} = \frac{347652000}{388920} = 89.4\%$$

b. The thickener Required Area

Using Talmage and Fitch method:

Mass flow rate of slurry (M) = 80 tph

Mass flow rate of solids in the slurry = % solids × M = 0.25 × 80 = 20 tph

The time associated with the critical point of the settling curve (tu) = 260
  s = 260/3600 h = 0.072222 h

The water to solid ratio (w/s) = 75/25 = 3

$$A = \frac{T}{G_{TEF}}$$

$$G_{TEF} = \left(\frac{H_F}{t_u}\right) \times C_F$$

$$C_F = \frac{w}{w/s} = \frac{1000}{3} = 333.333$$

$$G_{TEF} = \left(\frac{0.29}{0.072222}\right) \times 333.333$$

$$G_{TEF} = 4.015385 \times 333.333$$

$$G_{TEF} = 1338.46$$

But the mass flow rate of solids in the slurry T is 20,000 kg/h

$$A = \frac{20000}{1338.46} = 14.94\,m^2$$

References

1. Wills, B.A., and Napier-Munn, T.J. (2006): *Mineral Processing Technology*, 7th Edition. Amsterdam: Elsevier Science and Technology Books.
2. Eriez MACSALAB Filter Press (www.eriez.com/Images/Product-Images/Size-Reduction-Equipment/FilterPresses.jpg?Medium, Accessed 25th April, 2022).
3. Long, C.H., and Gruner, H. (2022): Drying Process, Material Processing, Britannica. In *Bricks and Tile Drying* (https://www.britannica.com>drying).
4. Savona Equipment (https://medium.com/@marianna_56839/thickeners-types-working-principle-applications-3d92a3725e8a, Accessed 3rd September, 2020).

5. Mettler Toledo: Flocculation: Theory and Background (2020): (www.mt.com/de/en/home/applications/L1_AutoChem_Applications/L2_ParticleProcessing/Formulation_Flocculation.html, Accessed 29th July, 2020).

6. McCabe, W.L., Smith, J., and Harriot, P. (1993): *Unit Operations of Chemical Engineering: Chemical Engineering Science*, 5th Edition, Vol. 6. New York: McGraw Hill, Inc. (http://doi.org/10.1016/0009-2509(57)85034-9, Accessed 3rd September, 2020).

7. 911 Metallurgist (2022a): *Difference Between Clarifier and Thickener* (www.911metallurgist.com/blog/difference-between-clarifier-and-thickener, Accessed 11th May, 2022).

8. 911 Metallurgist (2022b): *Thickener Control* (www.911metallurgist.com/blog/wp-content/uploads/2015/08/thickener_control.jpg, Accessed 11th May, 2022).

9. Romain, Hanna Lecture 12 MINE 292 (2012): *Stokes' Law and Settling Particles* (https://slideplayer.com/slide/3837883/, Accessed 17th August, 2020).

17 Automatic Control in Mineral Processing Plants

17.1 Introduction

A raw ore is treated in a sequence of coherent unit processes to obtain a final product, a specific mineral being predominant in the concentrate, from the ore body as received from the mine.[1] The typical processes to be controlled in a mineral processing plant are:

1. Mineral liberation which involves ore crushing, grinding and size separation to substantially release the mineral value from the associated gangue minerals in the ore
2. Mineral separation processes that use processes such as gravity, magnetic and froth flotation with the capability to select particles using their physical, physico-chemical surface and chemical properties
3. Peripheral processes such as feeding, thermal drying, conveying, slurry pumping, thickening, tailings disposal and mine backfilling.

The objective of the control procedure is to maximize the specific concentrate value while minimizing operating costs.

17.2 Principles of Automatic Control

Feedback control systems are used in industries. In a feedback control system, the functional relationships between the different elements are clearly indicated. It should be noted that the block diagram consisting of blocks and arrows represents the flow of control signals not energy through the system (Figure 17.1). The terms associated with the control loop are[2]:

a. The plant is the system in which the process takes place and by means of which a particular quantity or condition is subject to control.
b. The controllers are the control elements that generate the control signals to be applied to the plant.
c. The components to identify the functional relationship between the feedback signal and the controlled output is called the feedback elements.
d. The reference point is an external signal applied to the summing point of the control system to cause the plant to produce a specified action. This

DOI: 10.1201/9781003323433-17

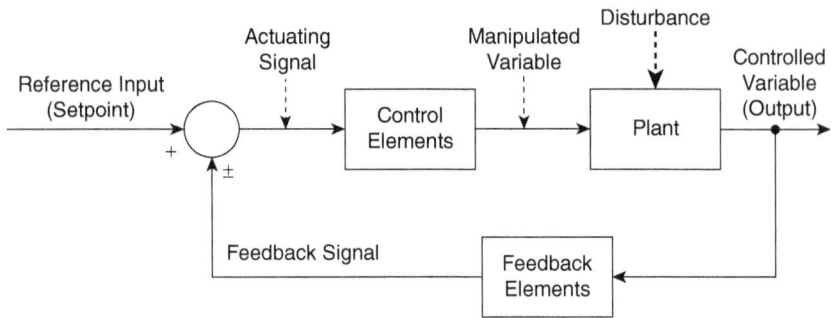

FIGURE 17.1 Feedback control system block diagram.

Source: **Courtesy of Technology Transfer (2020)**

signal is the desired value for a controlled variable and it is also called the "setpoint".

e. The quantity or condition of the plant that is to be controlled is called the controlled output. The signal stands for the control variable.

f. The actuating signal is obtained when the feedback signal which is a function of the output signal is relayed to the summing point and algebraically added to the reference input signal.

g. The actuating signal is also called the "error signal" and it represents the control of the control loop and it equals the algebraic sum of the reference input signal and the feedback signal.

h. The manipulated variable is the variable of the process acted upon to maintain the plant output, that is, the controlled variable at the desired value.

i. The disturbance is an undesirable input signal that upsets the value of the plant output.

Figure 17.1 presents the feedback control system block diagram.

17.3   Identification of Process Variables

The generic names of process variables can be classified as follows:

1. The input variables or independent or manipulated or control variables serve as inputs into the process, that is, the variables under which the ore is processed, with notation u.

2. The output variables are the final controlled variables and internal state variables which are other output variables apart from the controlled ones, with notations y and x, respectively. The variable x is process states that depend on control and disturbance inputs.

3. Disturbance variables, with notation d.

For a froth flotation process, the process variables for control to be considered are:

1. The manipulated variables are:
   a. Frother reagent concentration
   b. Collector reagent concentration
   c. Activator/depressant concentrations
   d. pH level
   e. Froth level
   f. Wash water quantity
2. The output final controlled variable y
   a. Concentrate grade
   b. Concentrate flow rate
3. The output variable x
   a. Stream compositions
   b. Stream flow rates
   c. Stream % solids
   d. Cell loads
   e. Froth loads
   f. Froth appearance
   g. Distribution of particle hydrophobicity
   h. Slurry to froth entrainment
4. The disturbance variables, d
   a. Ore feed rate
   b. Particle size and surface composition distribution

17.4   Process Modeling

The following are key points in modeling mineral processing plants:

1. There is a generalized control loop with tools developed based on process models
2. The loop tools comprise controllers, observers, reconciliation, predictors and optimizers
3. The type of model for each tool is different because each tool has a specific purpose
4. For instance, the control may be based on transfer matrices, the optimizer on phenomenological simulation, the predictor on Autoregressive—Moving—Average-Model (ARMA) equations
5. Traditional mathematical models can however be used to represent the processes

17.5   In-line Instrumentation

For the control of mineral processing operations, it is very important to have instruments to determine ore composition, slurry flow rate and particle size distribution in real time on properly obtained representative samples. The in-line ore composition determination is carried out with X-ray fluorescence analysis, while flow rate and slurry density are determined with magnetic flowmeter and gamma gauge densimeter, respectively. The coarse size particle size distribution is determined by processing the video image of the slurry, while the fine size range can be measured mechanically or deduced from laser diffraction or ultra-sound adsorption. Other properties such as slurry pH, motor power and pipe pressure are also required. The raw data obtained are used in the control loop with various tools to manipulate the process. The following should be noted regarding the instrumentation architecture:

1. The observer tool determines the key metallurgical variables and the economic performance index.
2. The observer performance is evaluated by the economic implication of the variance of its estimate from the set points.
3. The operating cost of the measurement system.
4. The robustness of the observer in terms of the accuracy of its estimates and ability to detect the control process' abnormal behavior.
5. The actuator refers to the communication hardware and software between the measuring devices and the data processing unit. It is the mechanism by which a control system acts upon the mineral processing operation 1

17.6   Overall Control Architecture

17.6.1   Control Objectives

Typical objectives for the control may be:

1. To produce a ground product with a specific percentage smaller than a given size obtained as a compromise between ore liberation and the cost of grinding.
2. In flotation circuits, it can be producing a concentrate at a given point on the grade-recovery curve.
3. For the overall grinding/separation process, the objective will be a concentrate on a particular combination of grade and recovery without the constraint of a product size specification.
4. The optimization of an objective function is based on some constraints. The objective function may be net revenue maximization or recovery per cent maximation at a constant grade and so on.
5. The constraints may be related to equipment limits, the safety of operation and environmental related legislation.

17.7   General Control Scheme

The following are the key points in the control scheme as shown in Figure 17.1:

1. The independent process and the disturbance variables (u, d) are used to run the processing operation

2. The concentrate's output of the processing in terms of x and y, as well as u and d and results of measuring devices, are received into the measurement block which then produces a vector of raw data feed into the fault detection block

3. The fault block indicates any problems with the process or measuring devices and resolves them by interaction with the Data Processing block (the process observers) before the observers' outputs are sent to the optimizer and controller

4. Data Processing block typically consists of the following devices:
   a. Signal filter is a device that removes some unwanted components or features or high frequency from a signal
   b. Data reconciliation tools are used in data migration to compare the target data against the original source data to ensure that the migration architecture has transferred the data correctly
   c. Soft sensor or virtual sensor refers to the software where several measurements are processed together or that deduce un-measured process variables from other variables to which they are correlated
   d. Image processors perform some operations on an image to obtain an enhanced image or to extract some useful metallurgical information from the image as delivered by video cameras

Based on a steady-state model of the processing plant, the optimizer will select for the recently obtained averaged operating conditions of the plant, set points that would maximize a metallurgical or economic-related objective function. The set points are delivered to the controller that uses the inputs based on its control model to adjust the u variables for the optimal operation of the process.

17.8   In-line Instrumentation for Automatic Control in Mineral Processing Plants

The automatic control of a mineral processing plant requires a continuous accurate measuring of the process stream. The advantages of automatic control include a reduction in the cost of operation and variation in product outputs as well as an increase in the grade of concentrate and plant productivity. The rapid development in process control is due to the availability of reliable and accurate real-time sensors and programmable logic circuits as well as microprocessors even in the harsh mineral plants' environment. The milestones in automatic control in the mineral processing plants include on-stream metal assaying of process streams at Outokumpu, XRF analysis of iron ore on moving conveyors, accurate ore weighing with an electronic weightometer, online moisture measurement, online particle size analysis in grinding circuits using camera and image analysis software. The real-time data enables the final control elements-the servo valves, pumps and variable speed motors to control the process variable.

In order to carry out automatic control in a mineral processing plant, instruments are connected to the conveyor belts and slurry conveying pipelines to obtain data that can be transmitted in real-time to the plant control computer system. The assays obtained in real time are used to control individual concentration loops and to optimize the overall plant operation. For instance, for the flotation of zinc sulphide from a complex ore, the rate of depressant addition is increased or decreased based on the set bounds for the assays of the concentrate and the middlings. The economic parameters derived from the assays of the feed, concentrate and tailings are used to effect the overall control of the flotation circuits. Automatic controls of the zinc flotation circuits were found to have curbed processing instability arising from disturbances and ore variations. The plant also recorded a reduction in cost due to reduced labour requirements during shifts and from the decreased frequency of assaying in the plant's laboratory. Examples of plants based on line or in-line or online instrumentation techniques include the following.[3,4]

17.8.1   A Mass Flow Meter

A mass flowmeter can accurately measure the mass flow of ore slurries that are typically difficult to determine by multiplying the volume and the mass flow per unit volume without restricting the slurry flow in the pipeline. The volume is determined by a magnetic flowmeter which works on the principle that a conductor moving in a magnetic field will cause a voltage to develop in it such that if the conductor maintains a constant dimension, the voltage induced will be directly proportional to its velocity. The meter is most appropriate for slurries carrying suspended solids and those with a turbulent flow because it works by an electrical averaging of increments in velocities in a flow profile that is not regular. Figure 17.2 shows the layout of a mass flow measuring system.[5]

17.8.2   The Magnetic Flow Meter

It determines the volumetric flow rate of a slurry and operates without any moving components. It can be used for a conductive liquid such as wastewater. The device generates a magnetic field through which the electrically conducting slurry in the pipeline flows and develops a voltage whose magnitude varies directly with its velocity through the pipeline according to Faraday's law. The pipeline wall is equipped with electrode sensors that detect the voltage developed and deliver it to an electronic system for processing to volts. Based on Faraday's Law, the signal voltage (E) developed depends on the average slurry velocity (V), the magnetic field strength (B) and the length of the conductor (D) which translates to the length between the two electrodes.

17.8.3   Ultrasonic Flow Meter

Ultrasonic flow meters use acoustic vibrations to determine the rate at which a liquid like wastewater or another dirty fluid that is conducting and having entrained bubbles, discontinuities of acoustic type or particles in suspension to reflect the ultrasonic signal flows. It works without contact with the liquid as it is clamped externally

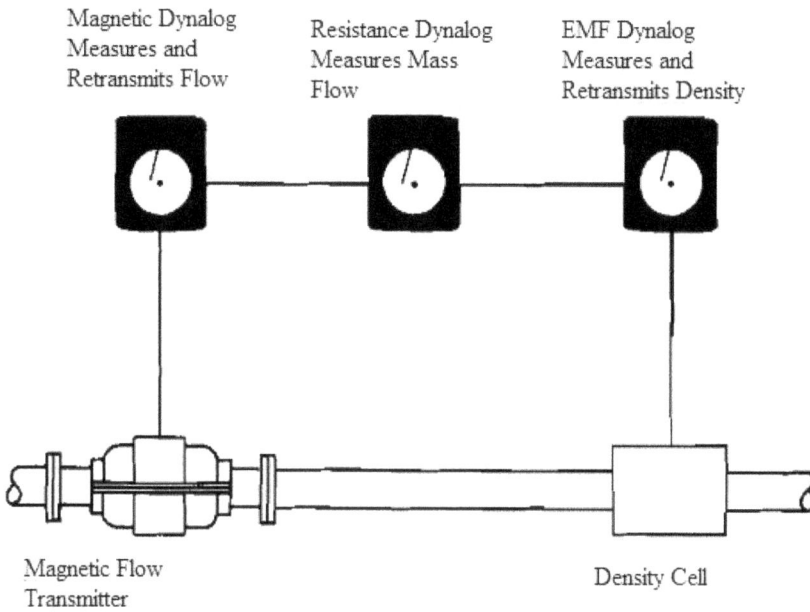

FIGURE 17.2   Mass flow measuring system.

Source: **Courtesy of 911 Metallurgist (2022a)**

to the body of the pipe carrying the liquid. There are two types-the Doppler and tran-
sient time. The Doppler effect refers to the phenomenon in which a sound appears to
have a shorter wavelength as its source approaches and vice versa. When an ultrasonic pulse is transmitted into the liquid, it will reflect after interacting with the
liquid's discontinuities with a change in frequency which is directly proportional to
the liquid's flow rate. Figure 17.3 is a schematic view of an ultrasonic flow meter.[6]

17.8.4   Nucleonic Density Gauge

This gauge enables the exact determination of the density of a slurry in real time.
In a processing plant, it can determine the densities for the underflow and overflow
streams to check that the float and separation tanks densities are correct. It is also
used to determine slurry density in coal washing to know the time coal middlings are
separated from the unwanted waste. The nucleonic gauge uses the resultant effect of
the ionizing radiation and matter interacting to carry out an analysis of the matter. It
can be used for static or continuous measurements and it operates with gamma rays
that penetrate into the target material through its container without any direct contact
so that its nature and properties remain unchanged. It is thus suitable for real time,
high-speed production and systems operating at elevated temperatures. The gauge,
transmission or backscatter type can use a radiation source or depend on the natural
radiation from the target material, respectively.

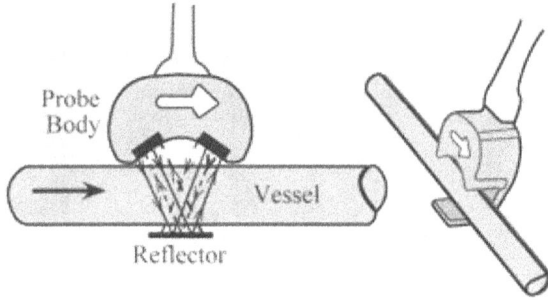

FIGURE 17.3   Schematic view of an ultrasonic flow meter.

Source: **Courtesy of Wikimedia (2022)**

FIGURE 17.4   A gamma ray density gauge.

Source: **Courtesy of 911 Metallurgist (2022b)**

In the backscatter operation, radiation originating in the target material is deflected by a shielding material in the gauge before it gets to the sensor for detection. The density of the material is directly proportional to the radiation detected. Figure 17.4 shows a gamma ray density gauge.[7]

17.8.5   Non-nuclear Density Gauge

The density of slurries is determined in real time by placing the density gauge in-line and the data can be used to control production and thus enhance processing performance. The traditional gamma ray density meters are being replaced with an electromechanical type such as the Alia density meter because of the risks associated with gamma radiations and the problem of obtaining a license for its installation. The density meter has an actuator that exerts a force with a known magnitude and frequency on the slurry flowing in the gauge while an accelerometer it also carries determine the resulting acceleration. Newton's second law is then used to determine the mass and volumetric flow rates knowing the volume of slurry involved.

17.8.6   On-line Chemical Analysis

The on-stream chemical analysis system typically has a source of radiation which is passed into the slurry to generate fluorescent radiations or photons characteristic of the elements in the slurry. The response photon radiations are taken into a detector that quantifies the amounts of each element represented by the characteristic radiations. For the in-stream probe systems, a sensor is placed in the slurry or near the slurry while the sample is excited using a low-energy source such as radioactive isotopes.

References

1. Hodoiun, D. (2009): *Automatic Control in Mineral Processing: An Overview.* IEACMMN, Vina del Mar, 14–16 October (https://www.sciencedirect.com/science/article/pii/S1474667016325678).
2. Technology Transfer (www.techtransfer.com/blog/basics-process-control-diagrams/, Accessed 2nd August, 2020).
3. Alia Instruments (2022): *Density Meter For Slurry | Simple, Reliable and Robust | Alia Instruments* (https://aliainstruments.com/adm-slurry-density-meter, Accessed 3rd March, 2022).
4. SRO Technology (2022): *Nuclear Density Gauge* (www.srotechnology.com/density-gauges, Accessed 6th May, 2022).
5. 911 Metallurgist (2022a): *Slurry Mass Flow Measurement* (www.911metallurgist.com/slurry-mass-flow-measurement/, Accessed 6th May, 2022).
6. Wikimedia (2022): *Ultrasonic Flowmeter* (https://upload.wikimedia.org/wikipedia/commons/1/12/Tttecnology.gif, Accessed 6th May, 2022) By No Machine-Readable Author Provided. Moraviaspy Assumed (Based on Copyright Claims). No Machine-Readable Source Provided. Own Work Assumed (Based on Copyright Claims), Public Domain (https://commons.wikimedia.org/w/index.php?curid=1859884).
7. 911 Metallurgist (2022b): *Specific Gravity to Density of Pulp or Slurry* (www.bing.com/images/blob?bcid=qDu3K1a4bSMERA, Accessed 6th May, 2022).

18 Introduction to Flowsheets Design

18.1 Introduction

The flowsheet is a diagrammatic presentation of the sequence of unit operations in the plant. Flowsheets can be presented in three ways, that is, as a block diagram, a simple line flowsheet and a process flow diagram/quantified flowsheet. The block diagram is the simplest form of flowsheet in which all similar operations are grouped together in a block from which arrows are directed to show the flow of subsequent operations. It consists of three blocks, namely, the comminution, the separation and the product handling as reported by Wills and Napier-Munn (2006)[1].

In the block diagram shown in Figure 18.1, the run-of-mine ore flows into the comminution block where crushing and grinding unit operations to obtain a substantial mineral value particles liberation from the gangue mineral particles take place. The separation block refers to the unit operations to be applied to the ore to obtain a concentrate consisting mostly of the mineral value and tailings containing mainly the gangue minerals. The product handling block has to do with unit operations such as sedimentation, thickening and thermal drying to obtain a concentrate largely free of moisture.

The simple line flowsheet is made up of a sequence of arrows that shows the flow of the processing operations through specific machines and undersize and oversize indicators for the ore flowing through the process. In the line flowsheet, the ore is crushed in stages to obtain the output suitable for grinding in the rod, ball and other mills depending on the degree of fineness required. After grinding, size classification is carried out before delivery to the most suitable separation technique that produces the concentrate and tailings. The (+) and (−) signs are used in the line flowsheet. The (+) sign refers to the oversized fraction of the ore material returned for further comminution treatment, while the (−) represents the undersized fraction of the ore material allowed to proceed to the next operation. A typical process flowsheet is a complex line flowsheet that uses standard symbols to represent different unit operations as shown in Figures 18.2 and 18.3. It also indicates the grade and recovery data for the concentrate and tailings at the end of each processing step.

The quantified flowsheets are line flowsheets showing more slurry operating details such as the % solids and slurry flow rate.

Figure 18.2 shows the flowsheet for the open and closed circuit processing of a slurry with a volumetric flow rate of 500 litres/hour.[2]

In the flowsheet for copper ore processing (Figure 18.3)[3], it can be seen that the ore delivered from the mine was subjected to gyratory primary crushing. It was after ground by semi-autogenous and ball milling for froth rougher flotation followed by regrinding and final cleaner flotation. The concentrate was then dewatered.

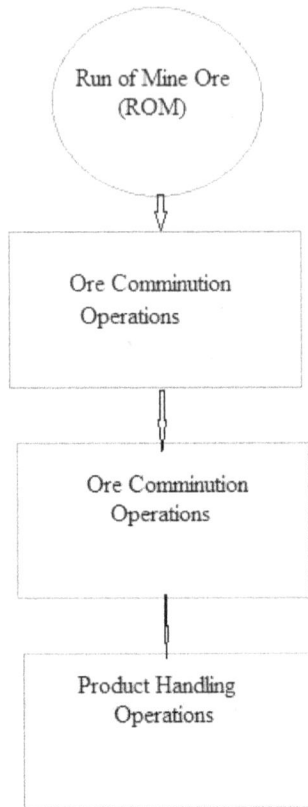

FIGURE 18.1   A simple block diagram.

18.2   Quantified Flow Data Derivation

A quantified flowsheet contains flow data that show the flow rate of the solids in the slurry, the % moisture in the slurry and the icons for each unit operation that constitutes the processing operations in the plant Stanley 1987.[4]

18.2.1   Worked Example

The moist solid feed rate into a ball mill-classifier circuit is 80 tph of crushed ore having 6% moisture. There are three hydrocyclones with each delivering 20 tph of dry underflow feeds diluted with water to water–solid ratio of 0.60 before entry into the ball mill. The total circuit feed was again diluted with water at the inlet of the ball mill to obtain a slurry with a water–solid ratio of 0.58 at the mill discharge. The mill discharge was further diluted to obtain a water–solid ratio of 1.3 for the cyclone feed. The cyclone overflow is the final product delivered for leaching at a water–solid ratio of 0.55. Construct a flow data table and a quantified flowsheet for the process streams

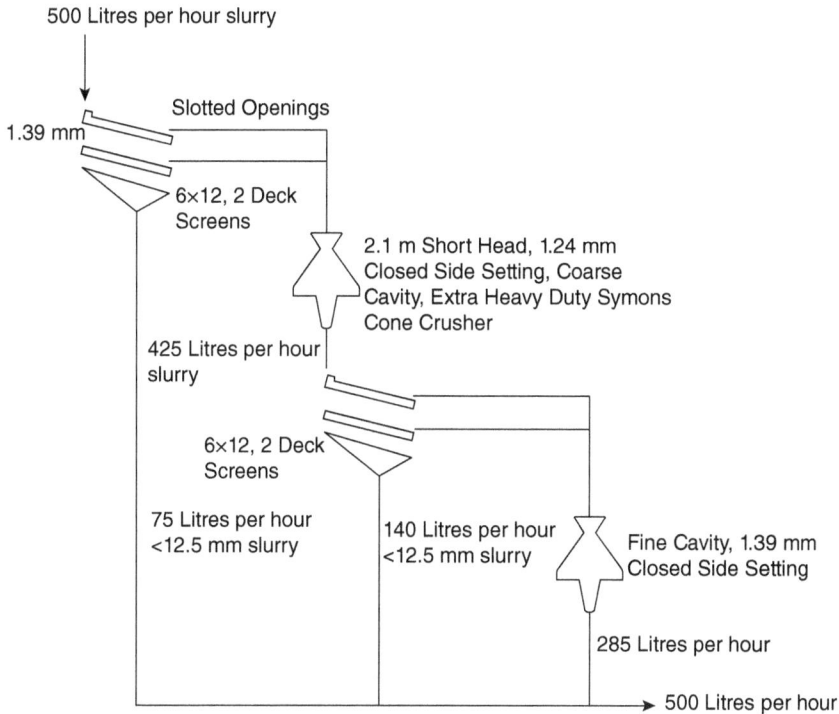

FIGURE 18.2   Open circuit cone crushing circuit.

Source: Courtesy of 911 Metallurgist (2022a), adapted

Solution

The problem is solved using the following sub-divisions:

Circuit feed

Ms = mass flow rate of dry solids charged by the conveyor = 80 tph
ML = mass flow rate of water in the moist solids = $6/94 \times 80 = 5.11$ tph
Mp = mass flow rate of the moist solids or pulp = $80 + 5.11 = 85.11$ tph

Cyclone underflow

Ms = mass flow rate of dry solids = $3 \times 20 = 60$ tph
ML = mass flow rate of water in the slurry = $60 \times 0.6 = 36$ tph
MP = $60 + 36 = 96$ tph pulp
% solids = $60/96 \times 100 = 62.5$
% Liquid = $100 - 62.5 = 37.5$

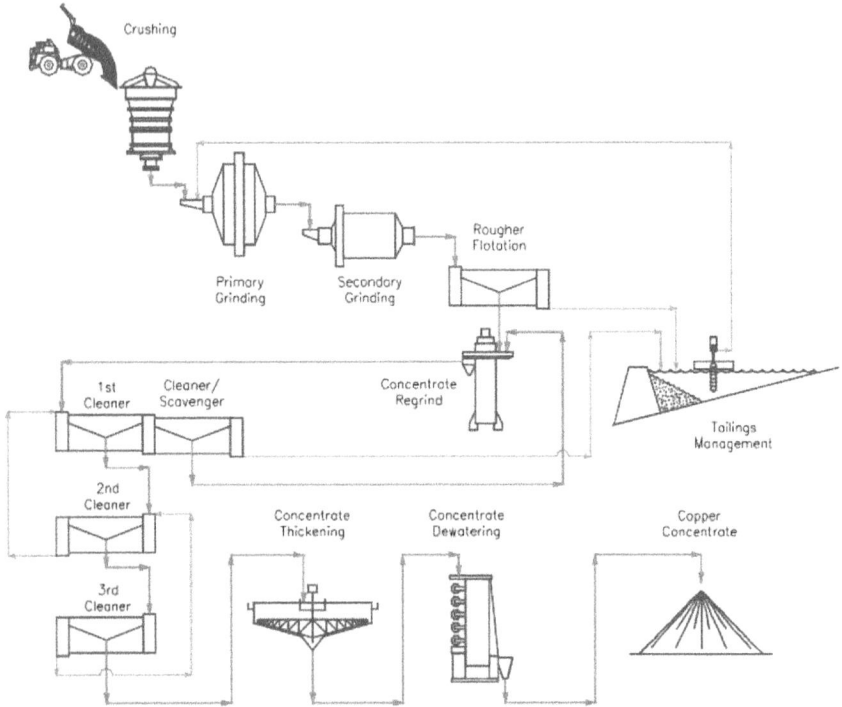

FIGURE 18.3   The copper ore processing flowsheet.

Source: **Courtesy of 911 Metallurgist (2022b)**

Ball mill discharge

$Ms = 80 + 60 = 140$ tph
$ML = 140 \times 0.58 = 81.2$ tph
$MP = 140 + 81.2 = 221.2$ tph
% solids $= 140/221.2 \times 100 = 63.29$ tph
% Liquid $= 100 - 63.29 = 36.71$
Water/solid $= 36.71/63.29 = 0.58$

Inlet dilution

$ML = ML$ at discharge sump $- MLiq$ (Circuit feed + cyclone U/F)
$= 81.2 - (5.11 + 36) = 40.09$ tph

Cyclone feed

Dilution ratio $= 1.3$
Mass flow rate of solids into the cyclone $= 140$ tph

Mass flow rate of water in the cyclone slurry feed = 140 × 1.3 = 182 tph
Mass flow rate of slurry into the cyclone feed = 140 + 182 = 322 tph
% solids in the cyclone feed = 140/322 × 100 = 43.48
% water = 100−43.48 = 56.52

Cyclone overflow

Mass flow rate of solids in cyclone overflow = 140−60 = 80 tph
Mass flow rate of water in cyclone overflow = 80 × 2.3 = 184 tph
Mass flow rate of slurry in the cyclone overflow = 80 + 184 = 264 tph
%solids in the cyclone overflow = 80/264 = 30.30
%water = 100−30.30 = 69.70
Mass of water for dilution in the cyclone overflow = 184−182 = 2 tph

TABLE 18.1
Flow Data Table

Stream	Description	Solids(tph)	Water tph (2)	Pulp (tph) (3)	%Solids
A	Circuit feed	80	5.11	85.11	94
B	Cyclone underflow	60	36	96	62.5
C	Mill inlet dilution	–	40.09	–	–
D	Mill discharge	140	81.2	221.2	63.29
E	Cyclone feed	140	182	264	30.30

The quantified flowsheet is presented in Figure 18.4.

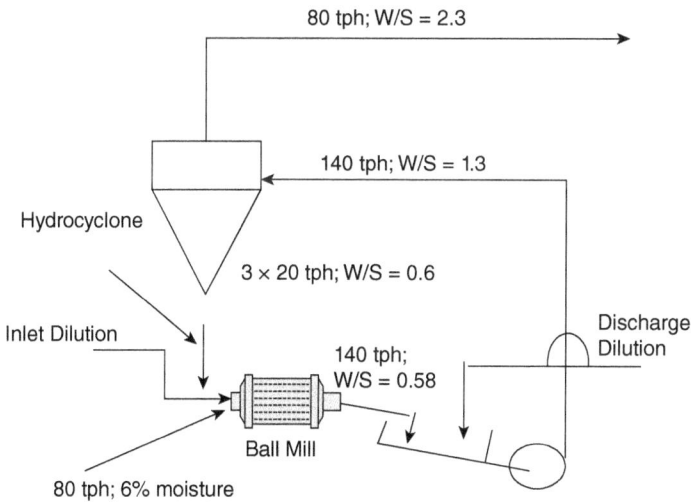

FIGURE 18.4 Quantified flow diagram with primary data.

References

1. Wills, B.A., and Napier-Munn, T.J. (2006): *Mineral Processing Technology*, 7th Edition. Amsterdam: Elsevier Science and Technology Books.
2. 911 Metallurgist (2022a): *Closed Grinding Circuits vs Open Grinding Circuits* (www.911metallurgist.com/blog/closed-grinding-circuits-vs-open-grinding-circuits, Accessed 8th June, 2020).
3. 911 Metallurgist (2022b): *Copper Process Flowsheet Example* (www.911metallurgist.com/blog/copper-process-flowsheet-example, Accessed 16th May, 2022).
4. Stanley, G.G. (1987): *The Extractive Metallurgy of Gold in South Africa*, volume 2. Johannesburg: The Soouth African Institute of Mining and Metallurgy (SAIMM) Monogram Series M7.

19 Assessment of Metallurgical Efficiency of Mineral Processing Operations

19.1 Parameters to Assess the Quality of Concentrates

When a run-of-mine ore is processed to obtain concentrates and tailings, it is always necessary to determine the quality of the concentrates obtained in comparison with the quality of the ore as received. Furthermore, it is also useful to determine the tailings' quality in terms of the mineral value content it holds as misplaced into it. The standard parameters to assess the quality of concentrates are grade, recovery, enrichment ratio and the ratio of concentration.

19.1.1 Grade or Assay of the Concentrate

The grade is the content of the end product that can be marketed in the material as either a feed, concentrate or tailings. Thus, in a metallic ore, the percentage of the metal value in the material as either the feed or concentrate or tailings is quoted as the grade or assay. Some metallic ores are assessed in terms of the percentage content of marketable oxide of the metal value of interest they contain. In such cases, the grade may be quoted in terms of the marketable oxide content. Examples are WO3 (tungsten trioxide or tungsten anhydride) and U3O8 (triuranium octaoxide). Tungsten trioxide is an intermediate product obtained after treating tungsten ores with alkalis, while triuranium octaoxide is the product of the crystalline oxidation of UO2 (urania) ore in the air at 250°C. For ores of very low grade such as gold metal, content may be expressed as part per million (ppm) or gram per ton (g/ton).[1]

For non-metallic ores, grade refers to the proportion of mineral content of interest, for example, the percentage of CaF2 in fluorite ores. Diamond ores are graded in carats per hundred tons (carat/100 tons), where 1 carat equals 0.2 g. Coal is graded based on its ash content, that is, the amount of incombustible mineral matter in the coal. The lower-grade steaming coals for power generation usually contain between 15% and 20% ash, while the much higher-grade coking coals (bituminous coals) used for metallurgical coke making in blast furnace iron-making are generally required to have an ash content of less than 12% plus the coking properties.

DOI: 10.1201/9781003323433-19

19.1.2 Recovery

For metallic ores, it is the percentage of the total metal contained in the ore feed that is recovered into the concentrate product. A recovery of 80% means that 80% of the metal in the ore feed is recovered into the concentrate product and 20% is misplaced into the tailings by-product. For nonmetallic ores, recovery referred to the percentage of the total mineral of interest contained in the ore that is recovered into the concentrate. This means that the recovery is usually expressed in terms of the valuable marketable end product. Non-metallic ores are generally associated with igneous/metamorphic rocks and cannot be economically smelted to produce metals such as iron, copper, gold, manganese and tin. Examples include coal, salt, sand, clay, fluorspar, limestone, marble, gypsum, natural gemstones, bitumen, peat, gravel and sulphur.

19.1.3 Ratio of Concentration (R)

The ratio of concentration is the ratio of the weight of the ore feed (heads) to the weight of the concentrates. Mathematically, $r = \dfrac{F}{C}$. The value of r generally increases with increasing grade.

19.1.4 The Enrichment Ratio (RC)

Enrichment ratio is the ratio of the grade of the concentrate (c) to the grade of the feed (f)

$$r_c = \frac{c}{f} \qquad (19.1)$$

This is also related to the efficiency of the process.

19.2 Relationship between Recovery and Grade

There is an approximate inverse relationship between recovery and grade in all separation processes. This results from the practical observation that an attempt to increase the grade of a concentrate leads to more metal value misplacement to the tailings implying a lower recovery in the concentrate. Conversely, an attempt for high recovery in the concentrate will result in more gangue minerals misplacement into the concentrate implying a lowering of the grade.

19.3 Metallurgical Efficiency of Separation Processes

The metallurgical efficiency of separation processes can be determined by the following:

1. Recovery—grade curve
2. Schulz separation efficiency equation
3. The net smelter return curve

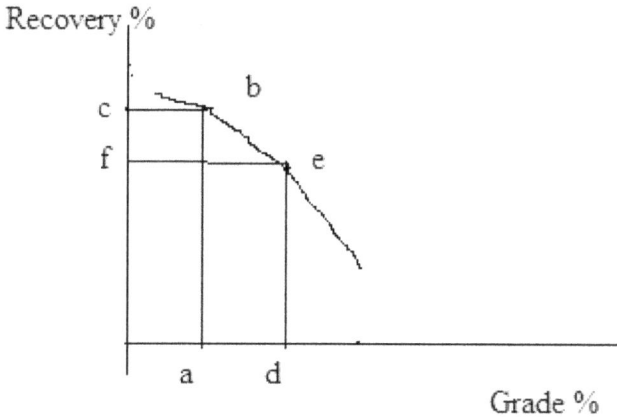

FIGURE 19.1   Typical recovery—grade curve.

Since recovery and grade are metallurgical factors, the metallurgical efficiency of separation] processes can be expressed by a curve (Figure 19.1) showing the recovery attained for any value of concentrate grade.

From the curve, it can be seen that as the grade increases from a % that gives a recovery of c% to d%, the recovery decreases to f% and vice versa. Mineral processing operations are carried out generally along a recovery-grade curve with a trade-off between recovery and grade such that the optimum point is determined by optimum profit. Since recovery and grade are inversely proportional, neither of them can be used independently to measure the efficiency of a metallurgical operation. When two processes are applied to a given ore and one of the two produces a concentrate with both a higher grade and a higher recovery, it can be uniquely determined that the process is the more efficient one. However, if one process yields a higher grade while the second produces a higher recovery, then one cannot uniquely determine the more efficient of the two processes. Consider the case given below:

Process A   Process B
Recovery -70 Recovery—50
Grade—45   Grade—60

In this case, we need another parameter (factor) that will combine both grade and recovery in its analysis to determine the metallurgical efficiency. Figure 19.2 shows the practical approach to improve flotation efficiency.[2] The processing operation should be carried out to push the recovery—grade curve forward as shown by improving on both recovery and grade simultaneously.

19.4   The Schulz Separation Efficiency

A single index of metallurgical efficiency or performance of a mineral processing operation was obtained by Schulz by combining recovery and concentrate grade parameters.

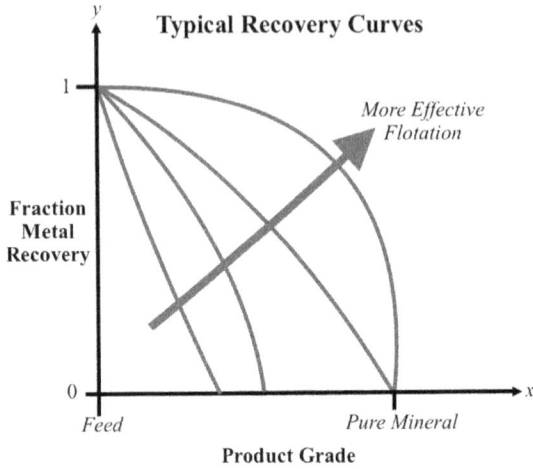

FIGURE 19.2 Shifts in the recovery-grade curve.

Source: **Courtesy of Wikimedia (2022)**

According to Schulz, separation efficiency (SE) is given by:

$$SE = R_m - R_g \qquad (19.2)$$

where Rm = % recovery of the valuable mineral into the concentrate
Rg = % recovery of the gangue mineral into the concentrate

$$Rm = \frac{100Cc}{Ff} = \frac{100C''c}{f}$$

Where F, C, f and c are the weights of the feed, concentrate, assay of the feed and assay of the concentrate, respectively. But

$$C' = \frac{C}{F}.$$

$$R_g = \frac{100C'(m-c)}{(m-f)}.$$

where m is the per cent metal content of the mineral value of an interest in its chemical state of occurrence, taken as pure

$$SE = R_m - R_g$$

$$SE = 100\frac{c'm(c-f)}{(m-f)f} \qquad (19.3)$$

19.5 Worked Examples

Example 19.1

A cassiterite ore that assayed 1.4% cassiterite as SnO_2 was processed to produce a concentrate that contains 46% cassiterite at a recovery of the valuable mineral of 68%.

Determine:

a. The % Sn in the ore
b. Calculate the Schulz separation efficiency for the process. Take atomic mass as: Sn = 119 g, O = 16 g

Solution

a. Determination of % Sn

Let the mass of the cassiterite ore = 100 kg

The mass of cassiterite in the ore $= \dfrac{1.4}{100} \times 100 = 1.4\,kg$

The molar mass of $SnO_2 = 119 + 2 \times 16 = 151$ g

The mass of Sn in the ore $= \dfrac{119}{151} \times 1.4 = 1.10$

Therefore, % Sn in the ore $= \dfrac{1.10}{100} \times 100 = 1.10$

b. For 46% grade and 68% recovery:

For the medium grade, the recovery is:

$$\text{Recovery} = 100\ C' \times \frac{c}{f}$$

where C' = the fraction of total feed weight that reports to the concentrate
c = concentrates' grade
f = feed's grade

$$C' = \frac{\text{Recovery} \times f}{100c}$$

Therefore,

$$C' = \frac{68 \times 1.4}{100 \times 46} = 0.0207 = 2.07 \times 10^{-2}$$

But Schulz separation efficiency is given by

$$S.E = Rm - Rg$$

$$Rm = \frac{100 C' c}{f} = \frac{100 \times 0.0207 \times 46}{1.4} = \frac{95.22}{1.4} = 68.01$$

Molar mass of the cassiterite Sn bearing compound = $119 + 2 \times 16 = 119 + 32 = 151$ g

Therefore $m = \dfrac{119}{151} \times 100 = 78.81$

$$R_g = \frac{100mC'(c-f)}{(m-f)} = \frac{100 \times 0.0207 \times 78.8\,(46-1.4)}{(78.81-1.4) \times 1.4} = \frac{7275.897}{108.374} = 0.88$$

SE = 68.01 − 0.88 = 67.14

Example 19.2

A copper concentrator treats a cuprite copper ore assaying 2.1% Cu and the two possible concentrates are:

Low-Grade: 42% Cu, 72% recovery
High-Grade: 54% Cu, 65% recovery

Determine:

a. % cuprite content of the ore
b. The Schulz separation efficiency for each of the two processes

Take atomic mass as: Cu = 64 g, O = 16 g

Solution

a. Determination of % Cu2O

Let the mass of the cuprite in the cuprite ore = 100 kg

The mass of Cu in the ore $= \dfrac{2.1}{100} \times 100 = 2.1$ kg

The molar mass of cuprite Cu2O = 64 × 2 + 16 = 144 g

The mass of Cu_2O in the ore $= \dfrac{144}{128} \times 2.1 = 2.4$

Therefore, %Cu_2O in the ore $= \dfrac{2.44}{100} \times 100 = 2.4$

b. For 42% grade and 72% recovery:

For the medium grade, the recovery is:

$$\text{Recovery} = 100\,C' \times \frac{c}{f}$$

where C′ = the fraction of total feed weight that reports to the concentrate
c = concentrates' grade
f = feed's grade in terms of cuprite

$$C' = \frac{\text{Recovery} \times f}{100c}$$

Therefore,

$$C' = \frac{72 \times 2.4}{100 \times 42} = 0.041143 = 4.11 \times 10^{-2}$$

Molar mass of the Cuprite ore the Cu bearing compound = 144 g

Therefore, $m = \frac{128}{144} \times 100 = 88.89$

But Schulz separation efficiency is given by

$$S.E = Rm - Rg$$

$$S.E. = \frac{100\, C'm(c-f)}{(m-f)f} = \frac{100 \times 0.041143 \times 88.89 \times (42-2.4)}{(78.81-2.4) \times 2.4}$$

$$= \frac{162.9257}{183.384} = 78.97\%$$

For 54% grade and 65% recovery:
For the medium grade, the recovery is:

$$\text{Recovery} = 100\, C' \times \frac{c}{f}$$

where C' = the fraction of total feed weight that reports to the concentrate
c = concentrates' grade
f = feed's grade in terms of cuprite

$$C' = \frac{\text{Recovery} \times f}{100c}$$

Therefore,

$$C' = \frac{65 \times 2.4}{100 \times 54} = 0.028889 = 2.89 \times 10^{-2}$$

Molar mass of the cuprite ore, the Cu bearing compound = 144 g

Therefore, $m = \frac{128}{144} \times 100 = 88.89$

But Schulz separation efficiency is given by

$$S.E = Rm - Rg$$

$$S.E. = \frac{100\, C'm(c-f)}{(m-f)f} = \frac{100 \times 0.028889 \times 88.89 \times (54-2.4)}{(88.89-2.4) \times 2.4}$$

$$= \frac{13250,54}{207.576} = 63.837\%$$

Therefore, the highest separation efficiency is obtained by the production of a low-grade concentrate, that is, the one with 42% Cu and 72% recovery.

19.6   The Net Smelter Return (NSR) Curve

The NSR expresses the economic factor for the separation efficiency. Although separation efficiency may be useful when plant performance at different conditions of mineral recovery and selectivity is to be compared, it does not take into account the critical economic factor, that is, profit. It has however been established that a high value of Schulz separation efficiency does not necessarily lead to the highest profit. Therefore, there is a need to determine the most economic combination of recovery and grade which will yield the highest profit per ton of the ore treated in the plant. The economic factors to be considered are:

1. The current price of the metal value of the interest in the ore and this is called valuation of the metal value
2. Transport cost of the ore concentrate to the smelter or the refinery
3. The cost of smelting treatment which depends much on the grade of the concentrate received
4. Penalty or bonus—the penalty is often imposed if the ore contains some undesirable gangue elements that adversely influence the smelting process. For example in the smelting of cassiterite to obtain tin, the presence of arsenic beyond a minimum level in the ore is penalized, while the content of precious metals such as gold and silver is awarded a bonus.

There will be a trade-off as explained here:

1. A concentrate with high grade will require a lower cost of smelting but the corresponding lower recovery of metals translates to lower profits on contained metal.
2. A concentrate having a low grade means a corresponding higher recovery and this will lead to higher smelter costs. In addition, it also translates to higher transport costs due to the higher inclusion of gangue minerals in it.

The net smelter return (NSR) can be determined for any grade and recovery combination using the following equations:

$$\text{NSR} = \text{Payment for the metal contained in the concentrate} + \text{Bonus} \left(\text{if any}\right)$$
$$- \left(\text{Smelter charges} + \text{transport Cost} + \text{Penalty} \left(\text{if any}\right)\right)$$

Figure 19.3 shows the variation of NSR with concentrate grade.[1]

Figure 19.3 shows that the NSR varies with concentrate grade. The optimum separation that will yield the highest NSR occurs at the maximum, P. The concentrate's grade corresponding to this point on the X-axis and its associated recovery obtainable from the recovery-grade curve will then be the target production combination.

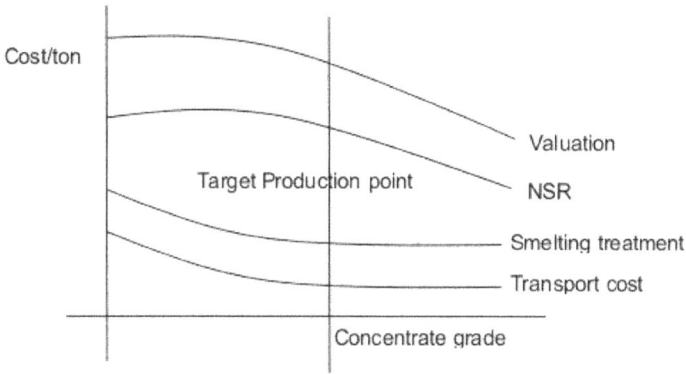

FIGURE 19.3   Variation of NSR with concentrate grade.

Source: **Wills and Napier-Munn (2006) (Courtesy of Elsevier Book Publisher, 2022)**

19.7   Practical Factors that Affect Metallurgical Efficiency

The practical factors that influence metallurgical efficiency include the following:

19.7.1   Liberation of Mineral Value

A major objective of comminution is to release or liberate mineral value or values from the associated gangue minerals at the particle size that is as large or coarse as possible. The successful liberation of mineral values at the particle size that is as large as possible has the following advantages:

(1) Energy use is reduced since the ore is not required to be ground to too much finer sizes before liberation is achieved
(2) The amount of ore fines produced is reduced
(3) It becomes easier and cheaper to carry out stages of separation that follow at the large size of the coarsely liberated ore

For high-grade solid product concentrate, good liberation at the coarse size is essential. However, for product concentrate to be subjected to hydrometallurgical processing such as leaching, it may be necessary to further grind the ore finer in order to properly expose the target mineral value to the reagent. In practice, however, it is hardly possible to achieve complete liberation even if the ore is ground smaller than the grain size of the mineral particles desired. The particles of the crushed ore will contain the mineral value either completely liberated or liberated with the gangue mineral to some degree. A collection of the particles of locked mineral value and gangue mineral particles is known as middlings.

The degree of liberation is defined as the proportion of the mineral value particles found as free fully liberated particles in relation to the content of the mineral value particles in the ore. If the degree of liberation is poor then the efficiency of mineral processing operations will also be poor.

19.7.2 Particle Size

It has been established that the smaller the particle size, the less efficient the mineral processing operation. Figure 1.8 in Wills and Napier-Munn (2006) shows the effect of particle size on the efficiency of different separation processes. For ores that are only liberated at ultra-fine sizes, that is, at sizes below 10 µm, it may become necessary to use the hydrometallurgy leaching to be able to effectively recover the mineral values.

References

1. Wills, B.A., and Napier-Munn, T.J. (2006): *Mineral Processing Technology*, 7th Edition. Amsterdam: Elsevier Science and Technology Books.
2. Wikimedia (2022): *Shifts in Recovery Curve* (https://upload.wikimedia.org/wikipedia/en/8/85/Shifts_in_Recovery_Curves.png, Accessed 23rd May, 2022).

20 Sustainable Mineral Processing

20.1 Introduction

There is a continual increase in demand for solid minerals and the metals extractable from them because of the growth of the world's economy and the emergence of new technologies. This situation necessitates sustainable processing of the depleting primary ore resources to minimize waste. Mineral resource utilization efficiency is, therefore, the key to sustainable processing of the primary solid mineral resources and a geometallurgy integrated approach is a pre-requisite in obtaining an improvement in the resource efficiency factor. The sustainable mining and processing of minerals will ensure that the present demand for minerals is met while guaranteeing future supply. The mining and mineral processing operations can be operated sustainably by addressing the following challenges.[1]

i. The problem of mineral raw materials wastage which can be minimized through proper waste management, recycling of tailings, reduction of processing losses and emissions.
ii. The challenge of energy consumption and the associated release of greenhouse gases into the environment. This can be addressed by reducing energy consumption.
iii. The problem of water shortage for mineral processing. This problem can be solved by reducing water consumption through the processing of ores at a high pulp density.
iv. The challenge of environmental pollution and damage resulting from mining operations. This challenge can be reduced through mine sites reclamation after mine closure.

The strategy for sustainable mineral processing is to improve the resource processing efficiency for a particular ore deposit resulting in the enhanced recovery of the contained mineral value with reduced water and energy consumption as well as lower waste loss and decreased environmental impact. Energy consumption can be classified as those for drilling, blasting, digging, ventilation, dewatering, diesel equipment usage, electric equipment operation, crushing, grinding and separation with grinding using diesel equipment accounting for most of the energy. Grinding is required for the sufficient liberation of mineral value particles which is the key pre-requisite for any ore processing method. The future processing of ore minerals has to contend with ores of lower grades, finer grains and complex mineralogy and thus requiring enhanced liberation and separation techniques to enable the output of concentrates with sufficiently high grades and recoveries.

DOI: 10.1201/9781003323433-20

20.2   Approaches for Sustainable Mineral Processing

Sustainable processing of minerals can be achieved by the following approaches.

20.2.1   The Application of Improved Ore Comminution

The effectiveness of ore comminution can be improved by ensuring adequate ore characterization in terms of ore texture and mineral associations that recognize liberation size variation in an ore body and thus guide against needless fine grindings to obtain target fine sizes for a specific mineral value. The breakage mechanism can also be adjusted based on a particular ore's texture to enhance mineral value liberation at coarser sizes and thus reduce energy consumption. The coarse size consist is also an advantage in the downstream separation processes. New mills in use in industries such as cement production like HPGR, stirred media mills and vertical spindle mills can also be used to reduce energy input more. Furthermore, comminution improvement can also be obtained through optimizing the grinding process circuit and plant capacity which may involve a correct choice of equipment.

20.2.2   Improved Ore Processing

In addition, a combination of comminution and physical separation can increase processing capacity as this will decrease the quantity of ore to be handled. Ore preconcentration that ensures the removal of gangue particles at coarser sizes using gravity concentrators, sensor-based sorters and other methods will also lead to a reduction in energy consumption. Resource utilization efficiency can also be increased by improving the recovery and selectivity of physical separations. For instance, the fine particles flotation process can be adjusted by adopting high-intensity dispersion and mixing of the particles, while coarse particles flotation are carried out at low intensity with large-sized bubbles.

20.2.3   The Geometallurgy Approach

Geometallurgy is an interdisciplinary technique designed to efficiently and sustainably exploit non-renewable mineral resources. It is intended to reduce the economic risk in the plant processing of an ore by ensuring a sound understanding of the ore, which is one of the two main variables that affect the efficiency of an ore processing. It combines the knowledge of geology and process mineralogy and uses data relating to mineral processing response to develop a spatially-based predictive model that enables plant operators to predict plant performance for ore inputs from a deposit into the future and until the mine is exhausted. It combines the traditional factors of mineralogy and rock type with the metallurgical characteristics that directly determine the way an ore responds during processing to define the identity of the ore.

Geometallurgy approach results in classifying metallurgical ore units into meaningful metallurgical types such that each unit may consist of a single ore type or different ore types having a unique set of compositional and textural properties based on which it can be predicted that the unit will respond similarly during processing. The efficiency of ore processing largely depends on the ore type and the operating philosophy, that is, how a plant is designed to be operated. The greater the geometallurgical understanding of the ore properties, the lower its effect on the efficiency becomes since its expected response during processing reduces in variation. Variation arising from poor geometallurgical evaluation leads to ore losses, change in processing response and dilution of the concentrate.

In geometallurgy, process models and simulations developed based on mineralogy and mineral processing tests on representative samples are used to optimize mineral processing operations. It is applied during ore exploration, mine planning, ore blending strategy, flowsheet design and management of tailings.

Process metallurgy relies on sound mineralogical characterization to study the characteristics of a rock or mineral at all scales—macroscopic, mesoscopic and microscopic, that is, over 1 m, 1 cm and 50 µm intervals, respectively. With more complex and refractory ores being increasingly mined, ore characterization that traditionally focuses mainly on in-situ ore tonnage and metal content grade determination now emphasizes cost efficiency and risk management which are provided by geometallurgy.

With geometallurgy, the following can be achieved:

i. Using sound economic assessment, aside from metal content grades, of the complete production chain from mining to the metallurgical processing to finished products for a typical variable ore deposit to define ore boundaries.

ii. Developing for valuable associated minerals. Profitable processing routes for co-production while ensuring the removal of gangue minerals before metallurgical extraction.

iii. The quantity and composition of tailings are also predicted by simulation while means of obtaining mineral values and/or reducing deleterious minerals from waste streams are provided.

iv. Enabling effective management of the environmental impact of ore processing.

v. It provides a solution to cost efficiency and risk management of ore deposits by deriving block models for the ore body to show how important metallurgical parameters are distributed to support process modeling. The knowledge of distribution also promotes process optimization and production planning. The information obtainable from block models include the following:

i. the knowledge of a rock's behavior from the preliminary drilling and blasting stages to the finishing stages of recovering metal and managing tailings

ii. a 4 D spatially oriented, predictive geometallurgical model which can be integrated over time unto a "virtual conveyor belt" for mine planning, ore blending, process optimization as well as flowsheet design optimization is derived by building the knowledge in (i) block by block

iii. the model in (ii) is further integrated with actual operational variables relating to the plant's finance, environment and risk analysis

iv. a decrease in the risks associated with the plant's operation by reducing avoidable metal losses to a minimum, effectively managing environmental issues and adopting variability in ore usage

Geometallurgy covers the fundamentals of mineral processing and the characterization of ore deposits. It is a comprehensive and quantified approach to characterizing ores taking critical processing parameters such as blasting, crushing, grinding, liberation, recovery of values and environmental preservation into consideration. The availability of knowledge derived from geometallurgy has led to an improvement in the economic optimization of mineral production and its sustainability, reduction in technical risk as well as better ability to make a forecast about future production, among others.

The particle-based geometallurgy approach treats various mineral sand particles as the common parameters passing through the geometallurgical program from the geological data collection to the final stage of process simulations. The approach comprises three quantitative models, that is, the geological, particle breakage and unit process models, representing the geological, comminution and separation stages, respectively. The geological model describes the modal composition and texture of the ore quantitatively and spatially. The particle breakage model describes in quantitative terms the kind of particles that will be produced as the rocks described by the geological model is crushed and ground. The unit process models describe quantitatively how broken particles will behave in different unit operations. The models are finally combined in a simulator that runs the process simulation and derives process performance parameters such as throughput, energy consumption, concentrate recovery and grade as well as tailing properties. for each ore block individually.[1]

In a geometallurgical program, efforts are made to produce a reliable model of an ore deposit and the mineral processing to treat the ore. A geometallurgical program development involves the following steps:

i. Geological data are collected by drilling through the ore body, applying drill core logging, taking relevant measurements and conducting chemical and mineralogical analyses, among others
ii. Determining places on the ore deposits where samples are preferably collected guided by geological data obtained
iii. Carrying out bench testing of the identified samples to obtain process model parameters
iv. Checking if the geological ore-type definitions are valid from a metallurgical perspective and developing from the geometallurgical domains, a new ore-type definition, if necessary
v. Estimate important metallurgical parameters from the geological database using derived mathematical relationships
vi. Derive a metallurgical model of the process comprising unit operations which operate using the metallurgical parameters defined earlier
vii. Simulating the plant using the metallurgical process model derived and the distributed metallurgical parameters as the data set
viii. Calibrating the models using benchmarking for existing operations

The benefits of the geometallurgical program in comparison to the traditional approach include the following:

i. The ore resources under consideration are used more economically because ore boundaries are defined on the basis of the metallurgical performance forecasted

ii. The process can be tuned based on the plant's ore feed information obtained before processing and hence better metallurgical performance can be achieved

iii. The more comprehensive knowledge of the ore body available through geo-metallurgy enables the mining to be better controlled

iv. Improvement in plant optimization becomes better because the changes in the feed entering the plant are small or, at least, its control is better

v. Better chances for new technological solutions because ore-derived problems are identified beforehand and research programs can focus on solving these

vi. Consequent upon increased knowledge of the ore body, the process as well as through a more controlled process chain, the risks in the operation are reduced

vii. The improved economical optimization of the full operation becomes possible when metal prices, alternative products and costs of commodities are considered

The particle-based geometallurgy approach uses mineral particles in samples to construct a link between geology and metallurgy, Therefore, the samples to be used for all the metallurgical tests must be truly representative of the bulk and the original ore deposit. The approach is made up of three models:

a. The geological model that provides the mineralogy by ore block quantitatively as against the traditional approach that yields qualitative and semi-quantitative data. The model provides a modal composition, that is, mineral content by weight percentage and information on texture throughout the ore body.

b. The particle breakage model which predicts the type of particles that will be produced when different ore blocks are comminuted.

c. The unit process model that predicts how different ore particles behave in a unit process.

The reliability of the geometallurgical programs is usually adversely affected by the inadequacy of the information obtained from the drill cores and the small size and number of samples that represent ore bulks in tons sent for metallurgical testing. The common practice is to carry out tests on about ten samples that are selected carefully and prepared, although less than ten samples are sometimes used. This emphasizes the need to place strict conditions on sampling and its preparation to reduce the error that may render the data obtained useless.[1]

It has been noted that new rock measurement and analysis techniques are required to make the sampling and definition of geometallurgical domains more reliable. To achieve this, a very large number of samples, ranging from thousands to hundreds have to be tested and need to involve techniques that are fast and not expensive. The techniques can be divided into three categories: techniques for measuring rock properties, techniques of quantitative mineralogy and techniques for geometallurgical tests.

A typical geometallurgical technique is the logging of petrophysical parameters directly from the drill core to assess rock properties, while quantitative mineralogy determines the mineralogical properties such as the modal composition of minerals, grain size of minerals, mineral associations as well as detection of some minerals. Techniques used for quantitative mineralogy include optical microscopy, image analysis and scanning electron microscopy. The methods for direct determination of metallurgical response are called geometallurgical tests. For comminution, the geometallurgical tests reported include rotary breakage and portable hardness tests. For the downstream ore processing, a bench-scale dry magnetic separator has been used to determine the metallurgical response of iron ore to magnetic separation, while a small-scale flotation test called JK mineral separability indicator was used for froth flotation.

In a full mineral processing operation, metallurgical or geometallurgical tests cannot directly produce a forecast of the behavior of a single ore block and this makes it necessary to build a mineral processing model. The mineral processing model consists of unit processes that make up the full ore processing circuit and consider static parameters like the number and size of the unit operations and the parameters relating to the ore which are equally geometallurgical parameters. Unlike in the traditional simulation practice where comminution and concentration models are separated, in geometallurgy, they are combined and for a steady state, simulation can produce metallurgical performances indices such as throughput, energy usage, recovery and grade of concentrate and properties of tailings for each feed blend or ore block.

A geometallurgical approach typically involves ore characterization and data modeling that are subsequently related to the modeling and simulation of downstream metallurgical extraction and refining processes. The application of the integrated approach of geometallurgy is a prerequisite to improving resource efficiency in mineral processing.

For geometallurgy, ore characterization has to include phase identification, modal mineralogy and mineral liberation and association based on automated mineralogy. The phase identification is carried out using optical microscopy, X-ray diffraction and scanning electron microscope. In addition, modal mineralogy and mineral liberation as well as studies of mineral associations are carried out. In optical microscopes, the color of minerals generated by selective absorption of certain light wavelengths in reflected or transmitted lights is an important diagnostic tool.[1]

20.3   Analytical Methods for Geometallurgy

20.3.1   The Transmitted Light Microscope

The transmitted light microscope is also known as the petrographic microscope. It is essentially made up of a white light source, a sub-stage condenser, an objective, an eyepiece and a stage for sample holding. It also has a polarizer that polarizes the light to vibrate in a single plane (i.e. plane-polarized light (PPL)), an analyzer, a second polarizer and a graduated rotating stage. The condenser serves to direct a cone of light on the thin section. The objective is a magnifying lens with the magnification power indicated. The analyzer is similar to the polarizer but it is oriented at right angles to the polarizer. When the analyzer is placed in the optical train the microscopy is said to have cross-polar. When the analyzer is not on the optical path, a bright image is obtained.

20.3.2   Description of Minerals in Thin Sections in Plane-polarized Light

Minerals can be identified by their natural body color in transmitted light, ranging from colorless (e.g., quartz and feldspar) to colored minerals such as biotite (brown), staurolite (yellow) and green (hornblende). The color obtained is due to the selective absorption of wavelengths in the visible white light which consists of radiations of wavelengths varying from violet to red. For minerals that appear colorless in thin sections, white light passes through the mineral and none of its wavelengths is absorbed, while opaque minerals such as metallic ores absorb all the wavelengths and appear black. For colored minerals, selective absorption of wavelengths occurs and the color seen in the transmitted light is a combination of the colors not absorbed. Some colored minerals exhibit pleochroism, that is, they change colors between two extremes when the microscope stage is rotated due to unequal absorption of light by the minerals in different orientations. The two extremes in color are each seen twice during a 360-degree rotation. Rock-forming silicates such as biotite and amphibole have this property.[2]

Habit, another property of minerals, refers to the shape a given mineral exhibits in different rock types. The shape may be acicular, when the crystal is needle-like fibrous when it resembles fibers or prismatic when it is elongated in one direction or platy/tabular when it is thin and flat.

Cleavage is another property by which minerals are identified. Most minerals can be cleaved or fractured along certain specific crystallographic directions which are related to planes of weakness in the mineral crystal structure. These planes are usually straight, parallel and spaced evenly in the minerals and have their crystallographic orientations denoted by Miller indices. Other properties under plane-polarized light are relief and alteration.

20.3.3   Description of Minerals in Thin Sections in Cross-polar Light

In cross-polar mode, all cubic minerals which are isotropic appear dark no matter their optical orientation. All other non-cubic minerals are anisotropic and exhibit colors that extinct or go dark four times during a complete rotation. For instance, quartz may vary from grey to white, while olivine may exhibit grey, blue, green or red colors. Other properties in the cross-polar view are twinning and zoning.

20.3.4   The Reflecting Microscope

Since most minerals are opaque, the study of minerals under vertically incident plane polarized light is called ore microscopy. Metallurgical microscopes that are also used to investigate metals and alloys are used. In a reflecting microscope, highly polished minerals are identified based on color, reflectivity, isotropism, anisotropy and hardness. The body colors observable in reflection mode are generally shades of grey and therefore color contrast between coexisting minerals provides more useful information. On rotating the stage, the color and reflectivity of isotropic minerals remain the same, while one or both anisotropic minerals change. In some opaque minerals, the incident light may pass through the surface and produce internal reflections from fractures and cleavages. When an analyzer is placed along the optical path, isotropic minerals will remain dark while anisotropic minerals will exhibit bi-reflectance which is known as birefringence for transparent minerals with four positions of extinction in a 360° rotation.

With the use of a low-power magnification and plane-polarized incident light, the following features will be observed in a reflecting microscope.

i. Transparent phases will appear dark grey in color. This is due to the fact that transparent minerals will typically reflect only about 3–15% of the light incident on them and transmit the remainder.
ii. On the other hand, opaque minerals appear grey to bright white as they typically absorb and reflect about 15–95% of the incident light. Some absorbing minerals that exhibit selective absorption appear colored.
iii. Specks of dust, holes, pits and cracks will appear black, while reflection from crystal faces in holes will appear as very bright patches.
iv. Scratches on a polished surface will appear as curves or straight lines.
v. Exposed surfaces under the polished surface will appear as bright patches.
vi. As a result of the different hardness of adjacent minerals, polishing relief that causes dark or light lines along grain boundaries will occur.

It should be recalled that metallic ores are rocks that contain the metal or its compound in sufficient amounts to make it economical to extract it.

20.3.5 Rietveld Analysis

This is an X-ray diffractometric analysis developed by Hugo Rietveld that derives a calculated diffraction pattern by refining factors such as lattice parameters, peak width and shape, as well as preferred orientation using a computationally intensive method. The derived pattern is then compared with that of an unknown and where both are nearly identical, the properties of the unknown sample such as crystallite size and size occupancy factors can be obtained.

20.3.6 Quantitative Evaluation of Minerals by Scanning Electron Microscopy (QEMSCAN)

The use of a conventional microscope for the examination of concentrate and tailings samples in thin and polished sections has been used to obtain valuable information that indicates the efficiency of mineral value liberation and mineral processing operations. It has a special use in probing processing problems relating to insufficient liberation of mineral values. However, new techniques of automated mineral analyses such as Mineral Liberation Analyzer (MLA) and Quantitative Evaluation of Minerals by Scanning Electron Microscopy (QEMSCAN) are increasingly used.[3, 4]

20.3.7   Mineral Liberation Analysis (MLA)

The Mineral Liberation Analyzer consists of an FEI Quanta 650F Field Emission Scanning Electron Microscope. It is equipped with two Bruker Quantax X-flash 5030 Energy Dispersive X-ray detectors (EDX). For the analysis, 3 g of the mineral sample particles are mixed with the same volume of graphite and epoxy resin to produce grain mounts. The grain mounts are polished and carbon coated using a vacuum evaporator. Mineral grains are identified from the Back Scattered Electrons (BSE) image and from the EDX-spectra of particles and grains distinguished in BSE imaging mode. MLA Software is then used to process and evaluate the data obtained in a specific measurement mode such as GXMAP. The GXMAP mode combines X-ray mapping with the XBSE image processing steps of BSE image acquisition, particulation and segmentation before X-ray spectra collection.

QEMSCAN is a custom-made automated mineral analyzer for the mineral industry. A typical QEMSCAN uses a field emission gun scanning electron microscope and is equipped with a high resolution Back Scattered Electron (BSE) system, Bruker Energy Dispersive Spectrometers (EDS) and a Spectral Analysis Engine (SAE) to analyse mineral phases. The measurements obtainable with QEMSCAN includes Bulk Mineralogical Analysis (BMA), Particle Mineralogical Analysis (PMA), Specific Mineral Search (SMS), Trace Mineral Search (TMS) and Field Scan. The information obtained from the measurement types mentioned are bulk mineralogy of samples, ore deportment, estimated grain and particle size, particle images, grain and particle shape, degree of mineral liberation, mineral association and theoretical grade recovery curve.

For sample preparation, the representative sample obtained by splitting and rifling is mixed with graphite and mounted in resin and then allowed to cool in 30 mm round holes. The sample blocks are carefully polished to a 1 μm diamond finish. Carbon coating is then applied to the sample. Figure 20.1 presents the QEMSCAN image of a fluvial sandstone.[4]

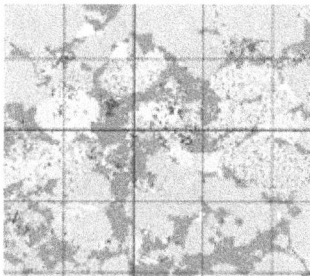

FIGURE 20.1   QEMSCAN image of a fluvial sandstone, grid size = 500 μm.

Source: Courtesy of Wikimedia (2022)

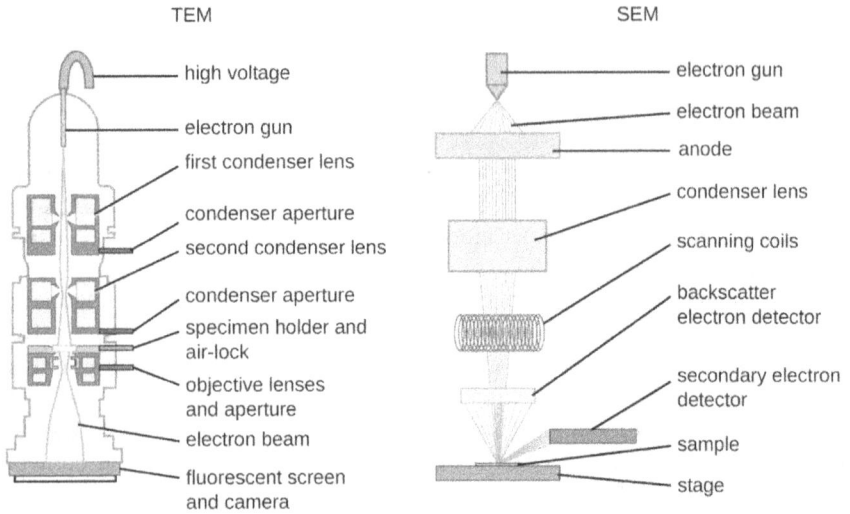

FIGURE 20.2   Basic components of scanning and transmission electron microscopes (pinterest.com, 2022).

20.3.8   Scanning Electron Microscopy (SEM)

The analysis of the image data obtained from a scanning electron microscope can be used to evaluate the efficiency of a mineral processing process. SEM-based image analysis has been used to characterize minerals in detail. MLA provides the percentage distribution of the mineral groups as liberated after comminution such as silicates, phyllosilicates, iron oxides, fluorites, sulphides and cassiterites. The silicate group may consist of quartz, garnet, feldspar and amphiboles, the phyllosilicate may have mica and chlorite. The sulphides may comprise galena, sphalerites, pyrite and arsenopyrite while the iron oxides refer to hematite and magnetite. Figure 20.2 shows the basic components of both scanning and transmission electron microscopes.[5]

20.3.9   Sustainable Circularity in Mineral Processing

In a circular economy, the useful life of materials and products which circulate in the economy is extended by creating loops that ensure their repeated use and repair as well as re-manufacturing. It is designed to reduce resource input for manufacturing and associated wastes and polluting emissions produced. The aim of the concept is to attain maximum efficiency in the use of finite non-renewable resources, recover residual values in expired materials and products as well as achieve a gradual shift to renewable substitutes for the finite non-renewable resources circulating in the economy. It is an economic system that is restorative and regenerative in practice. It is an alternative to the linear economy based on the "make-dispose" model and it ensures high sustainability of processing operations without reducing the profit level of the business or the number of its outputs into the economy.

The circular economy is based on three principles, that is, minimization of wastes and pollution, the extension of the useful life of materials and products as well as regeneration of natural systems. It distinguishes between the technological cycles involving finite resources and biological cycles that deal with how renewable resources are managed. In a circular economy, a product's lifetime can be extended practically by designing it to be easy to repair, more durable, promote its repair instead of replacement and manage its waste better.

In concept, a circular economy is a system designed to carry out the production and consumption of materials towards zero waste. It ensures the use, reuse and recycling of products without eventual disposal at any point of the product lifecycle.

The circular economy concept has been applied in mineral and materials processing as follows:

i. The recycling of critical and depletable rare earth metals obtained from their scarce primary deposits and ensuring their improved sustainable use as they circulate in the economy. For example, yttrium primary ore is predominant in a particular area of the world and it is considered a critical raw material (CRM). It has been reported that less than 1% of rare earth elements (REEs) used were recycled as of 2019 and recycling can only be economically viable if the market price per kg exceeded €9.54 per kg for yttrium as an example.

ii. The environmental benefits of recycling over-extraction from ore minerals are also considered in the circular economy of REEs as it has been established that recycling has fewer adverse effects on the environment than primary extraction. Thus, the circular economy may also need policy instruments that will promote recycling CRMs over primary production for increased profits and environmental safety.

iii. Efforts are made to reduce the waste of raw materials, optimize processing efficiency, reduce energy use and promote waste-free processing.

iv. Materials are being re-designed to enable them to be re-purposed, re-used and recycled as a transition from the linear economy dependent on finite resources of mining and petroleum to the circular that will use novel and renewable materials.

v. Development of new recycling methods for REEs that will ensure its preferred use over primary extractions.

vi. Development of self-healing materials such as the self-healing polymer coatings for pipelines in adverse environments.

vii. Addition of values to waste by-products to generate new useful products such as the conversion of the cotton gin waste into bio-degradable plastics.

viii. Development of innovative techniques to recycle valuable metals from their wastes as is being pursued for titanium alloy wastes.

ix. The emergence of new and functional business diversity in the mineral processing industry.

x. In mineral processing, it addresses the challenges which include; resource depletion and environmental pollution. The mining and subsequent mineral processing operations leave large volumes of waste in the form of waste rocks, emissions, tailings and wastewater, and the circular economy offer alternative options to handle these wastes. Precious metal such as gold which is becoming increasingly difficult to realize from primary sources has found circularity useful. The waste of electronic appliances contains gold in reasonable quantities and these can be efficiently recovered by leaching technology and target metal adsorption using bio-based adsorbent for an economical bio-friendly means of recycling the gold. Tailings are also processed using the same approach.[6,7,8]

References

1. Lamberg, P. (2011): *Geometallurgy-A Tool for Better Resource Efficiency* (www.researchgate.net/publication/282152217, Accessed 23rd March, 2022).
2. Munyao, N.C. (2014): *SGL 201-Principles of Mineralogy* (https://profiles.uonbi.ac.ke/cnyamai/classes/sgl-201-principles-mineralogy, Accessed 23rd March, 2022).
3. Centre for Materials Research (CMR) (2022): *QEMSCAN* (www.cmr.uct.ac.za/cmr/ra/process-mineralogy-qemscan, Accessed 23rd March, 2022).
4. Wikimedia (2022): *QEMSCAN* https/en.wikipedia.org/wiki/QEMSCAN#:~:text=QEMSCAN%20is%20the%20name%20for%20an%20integrated%20automated, registered%20trademark%20owned%20by%20FEI%20Company%20since%202009, Accessed 28th May, 2022).
5. Pinterest.com (2022): *Scanning and Transmission Electron Microscopes* (https://th.bing.com/th/id/R.3dbc70e645aa0d01104ac899f61370d8?rik=kbou1TRFtjbqxw&pid=ImgRaw&r=0&sres=1&sresct=1, Accessed 23rd May, 2022).
6. Corporate Finance Institute (CFI) (2022): *Circular Economy—Overview, Principles, Types of Cycles* (www.corporatefinanceinstitute.com, Accessed 6th January, 2022).
7. Institute for Frontier Materials (IFM) (2022): *Re-Designing Materials for a Circular Economy.* Institute for Frontier Materials (www.deakin.edu.au, Accessed 7th January, 2022).
8. Pomykala, R., and Tora, B. (2017): Circular Economy in Mineral Processing. *Mineral Processing E35 Web Conferences*, 18, 01024.

21 Processing of Rare Earth Minerals

21.1 Introduction

Rare earth elements (REEs) refer to the fifteen elements of the lanthanide series as well as yttrium and scandium, and these may occur in more than 250 different minerals. The REE deposits are found as carbonates, pegmatites and as placers but they are most often considered too lean in REE values for economic recovery. In view of this, only a few REE commercial deposits are available all over the world. These elements are used in many applications such as in the production of permanent magnets with high magnetic field strength, petroleum refining catalysts, additives for metal and glass and phosphors used in electronic displays as well as in defence products.

Green economy requires the use of electric generators and motors with high working efficiency but requires very strong permanent magnets such as iron-neodymium-boron (Fe-Nd-B) magnets which are currently rated to possess the highest magnetic strength. Fe-Nd-B magnets are alloyed with Dy to improve their high-temperature tolerance. It has been reported that samarium-cobalt magnets are more corrosion-resistant than Fe-Nd-B magnets.

Rare Earth Elements (REE) ores are found as solid rocks or as un-consolidated sediments comprising several REE minerals such as monazite, xenotime, bastnasite, steenstrupine and eudialyte in association with undesirable gangue minerals like quartz, calcite or feldspar. The mining of REE ores may be done by a pit or underground mining method based on the form of its occurrence and the location of the ore body. However, bastnäsite, monazite, and xenotime are the only REE-bearing minerals which have been mined commercially and they constitute over 95% of the world's supply of REEs. The chemical formula of monazite, [Ce, La, Nd, Th)(PO$_4$, SiO$_4$] indicates that cerium, lanthanum, neodymium and thorium can all substitute for one another in the mineral structure and silica can also substitute for phosphate.

Zhang et al. (2020)[1] reported that REEs are also recovered from coal and coal-based materials such as coal waste, acid mine drainage and by-products such as fly ash obtained by the combustion of coal. These materials are finely ground before being subjected to preliminary leaching and bio-leaching. Monazite with a density that ranges from 4.6 to 5.7 g/cm^3 is a reddish-brown phosphate mineral and it is an important ore for thorium, lanthanum and cerium. Monazite is a widely distributed mineral and occurs as an accessory mineral in granitic igneous rocks. A minor primary mineral is one not required in the naming of the rock. It is usually found as small isolated crystals with typical Mohs hardness that varies from 5.0 to 5.5. Other common minor minerals considered as accessories include topaz, zircon, fluorite, garnet, ilmenite, rutile, allanite, corundum, magnetite and tourmaline. A mineral is

DOI: 10.1201/9781003323433-21

considered an essential mineral in an igneous rock if it is a major primary mineral whose occurrence is essential to define the root name of the rock.

Xenotime with a density that ranges from 4.4 to 5.1 and monazite are the two major REE phosphate minerals and are also called heavy minerals because of their high densities which are greater than 2.9 g/cm^3. Xenotime is mainly used as a source of yttrium and heavy lanthanide metals. The chemical formula is YPO$_4$ in repeating units. The mineral usually occurs in brown or yellow color and as tetragonal crystals and rolled grains. Xenotime has been reported as the largest source of heavy rare earth elements (HREE).

The major mines for monazite and xenotime are found along the coasts of Southern India, Madagascar and South Africa. They have been reported to host large deposits of monazite sands. Most REE minerals have their origin in granites and gneisses and hence the concentrations of ThO$_2$ in monazite can range from 2% to 10% rendering the ore too radioactive for handling and storage. Xenotime is usually obtained as a by-product of monazite processing and hence it can be processed in a way similar to monazite, Figure 21.1 shows a monazite mineral.[2, 3]

Bastnasite (La(Ce)CO$_3$F) is a yellowish to brownish mineral that hosts rare earth elements, particularly cerium, a ductile grey metallic element of the lanthanide series

FIGURE 21.1   The monazite mineral.

Source: **Courtesy of Wikimedia (2022a)**

FIGURE 21.2   Bastanasite from Burundi.

Source: Courtesy of Wikimedia (2022b)

which are 15 metallic chemical elements with atomic numbers 57–71 from lanthanum to lutetium. These 15 elements and the chemically similar yttrium and scandium are referred to as rare earth elements. In addition to similarity in chemical behavior, scandium and yttrium also exhibit the tendency to occur in the same ore deposits as the lanthanides. The REEs are used diversely in industrial processes and the making of electrical and electronic components, lasers as well as glass magnetic materials. Figure 21.2 presents a bastanasite from Burundi.[4]

Zhang et al. (2020)[1] reported that coal-related materials such as coal, coal waste and by-products of coal combustion like fly ash have been processed to recover REEs. Physical beneficiation, acid leaching, ion-exchange purification, bio-leaching, thermal treatment, alkali treatment, solvent extraction, and other methods to recover metal values have been assessed and found to yield recoveries based on the properties of the feed material. Generally, physical beneficiation was noted to be a cheap option suitable to upgrade the feedstock as a preliminary stage. However, most studies gave extremely low recoveries except for when the feeds were first subjected to ultrafine grinding. This finding is largely due to the combined effects of complex REE mineralogy in the coal-based materials and small particle sizes of the feed. In the alternative, direct acid chemical extraction produced moderate recovery of values and the introduction of leaching additives, alkaline and/or thermal pre-treatment, greatly enhances the leaching processing.

21.2   Beneficiation Methods for REE Minerals

The beneficiation methods for REE ores typically vary depending on the differing characteristics of the deposit such as the grain size, mineralogy and texture of the ore and thus each ore has to be carefully examined to determine the appropriate processing route. Ores that are coarsely liberated are typically crushed and concentrated using physical methods such as gravity or magnetic separation. However, the finer-grained rare earth minerals need to be separated using a physico-chemical method like froth flotation, while the complex ores will require direct leaching after preliminary preparations to recover the rare earth elements.

The typical steps to determine a beneficiation approach are:

1. Collecting a representative sample of the ore
2. Quantitative mineralogy of the ore as received
3. Bench-scale trial of a range of mineral processing methods
4. Bench-scale optimization of the best method or a combination of methods
5. Locked cycle test work
6. Demonstration or pilot plant test work on tons of ore
7. Analysis of tailings for sulphide minerals and radioactive elements that can cause acid mine drainage and environmental damages, respectively

EURARE (2017)[5] reported the pilot plant processing of 3 tons of Greenland Kvanefjeld REE sample that assayed 1.3% rare earth ore (REO) and produced 250 kg REE concentrates that assayed 15% REO after the treatment of the ore in a pilot plant. The REE concentrate was then processed to obtain 25 kg of mixed REE carbonates. The REE mixed carbonates were further processed by solvent extraction to produce La, Ce, Nd and Pr chlorides. The rare earth elements can be subdivided into the Light and Heavy REEs (LREE and HREE). The LREE includes neodymium (Nd) and praseodynium (Pr), while HREE includes dysprosium (Dy). LREE are found in the mineral bastanasite and monazite, while HREE is hosted by xenotime and yttrium phosphate. However, the three minerals mentioned are not readily amenable to leaching and are associated with high content of thorium. REE minerals are paramagnetic and it is possible to separate them from the gangue minerals such as quartz and feldspar that are diamagnetic.

21.3 Pyrometallurgy and Hydrometallurgy in REE Processing

The ore minerals that host REEs need to be concentrated prior to the extraction of the metals. The beneficiation methods for REE ores are typically determined based on the differing properties of the deposit such as the grain size, and mineralogical and textural characteristics of the ore and thus each ore has to be carefully examined to determine the appropriate processing route. Ores that are coarsely liberated are typically crushed and concentrated using methods such as gravity or magnetic separation based on the physical properties of the ore. However, the rare earth minerals that require finer grinding need to be concentrated using a physico-chemical method like froth flotation while the complex ores will require direct leaching after preliminary preparations to recover the rare earth elements.

Pyrometallurgy and hydrometallurgy are the traditional routes to extract REE from its ores. Pyrometallurgy treats high-grade REE ore minerals by calcining them with a mixture of flux and reducing agents. The main disadvantages of the pyrometallurgy route are:

a. High energy consumption for slagging and reduction
b. The release of large volumes of greenhouse gases to the environment
c. During the calcining, the constituent REEs may be separated into different slag phases making it necessary to use hydrometallurgy to recover them

21.4 Hydrometallurgy

In view of its lower energy consumption and suitability to treat low-grade ores, hydrometallurgy is preferred to pyrometallurgy in the recovery of REEs from REE ore minerals and their concentrates. A concentrate of the rare earth element (REE)-bearing minerals bastnäsite and monazite was produced by a solid-stabilized Pickering emulsification process from an ore associated with gangue minerals such as dolomite, calcite, ankerite, silicate, quartz, pyrite, mica, hematite and barite. The REE-bearing minerals released exhibit better attraction for oil droplets than the gangue minerals due to differences in the tendency of the surfaces to get wet and finally adsorb to the oil–water interface while the gangue remains in the water phase. The enrichment ratio was determined as 2.9, with a recovery that exceeds 50%, and with no need to adjust the pH and/or apply a modifier for the surface.

An emulsion is a mixture of two or more liquids in which one of the liquids occurs as droplets of microscopic or ultra-microscopic size dispersed in the other liquid. They are produced from the liquid components either by spontaneous action or more frequently mechanically, by means such as agitated mixing, on the condition that the mixing liquids are not mutually soluble or are only so to a very limited extent. Emulsions are thermodynamically unstable and are stabilized by using agents such as bentonite or colloidal carbon or by forming films on the surface of the droplets using soap molecules. When emulsion-remain unstable, they will finally divide into separate liquid layers. Emulsions that are stable can be destabilized by freezing, heating or by adding a suitable third substance. Examples of regular emulsions include milk, which results from fat droplets dispersed in an aqueous solution and butter, formed when droplets of an aqueous solution get dispersed in fat. Other regular emulsions include cream, gels, pastes and vaccines. Emulsions stabilized by adding solid particles are called Pickering emulsions. They are resistant to coalescence and thus exhibit long-term stability.[6, 7]

Hydrometallurgy has been reported to have the following main disadvantages:

a. The intensive use of reagents-mineral acids and organic solvents
b. The output of secondary waste streams that are considered potentially hazardous
c. The complex process flowsheet for the extraction

21.4.1   Leaching of Eudialyte

Eudialyte, a complex REE, has been reported to be less abundant than xenotime, but it is more amenable to leaching treatment. It is an alkaline zircon silicate and can be reached at a pH of < 3. It has a long formula:

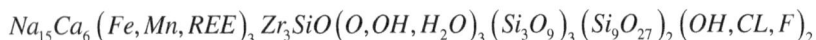

$$Na_{15}Ca_6\left(Fe,Mn,REE\right)_3 Zr_3 SiO\left(O,OH,H_2O\right)_3\left(Si_3O_9\right)_3\left(Si_9O_{27}\right)_2\left(OH,CL,F\right)_2$$

From the chemical formula, it can be seen that at least three silicates can form from the ore during leaching. Based on the pH range required, both organic and mineral acids can be used to leach the mineral. However, the leaching agent in addition to dissolving the HREEs will also dissolve the silicate associated with the ore. The dissolved silicate, called silicic acid, will later transform into gels that will hinder further chemical reactions.

Ore concentrates with grades ranging from 1 to 10% can be produced from the eudialyte ore on magnetic concentration treatments. To produce very high-grade concentrates assaying greater than 30% REE will require flotation after magnetic pre-concentration. In order to avoid the formation of silicates during leaching, the ore concentrate can be subjected to a carbochlorination process that converts Fe, Zr and Si that has an affinity for chlorine to volatile chlorides such as $FeCl_3$, $ZrCl_4$ and $SiCl_4$ that can be removed by distillation and fractionation. The remaining elements are then leached to produce soluble chlorides. A batch process in which 5% carbon was reacted in combination with chlorine and the solid ore particles in a crucible has been reported to produce REE yields higher than 30% at a reaction temperature of 750°C. It was suggested that a fluidized bed can replace the crucible for the batch process for improved efficiency in the process. The conventional leaching methods typically require the use of large quantities of acids and organic solvents and hence large quantities of water that pose hazards to the environment.[8]

21.5   Secondary Resources of REEs

For the sustainable supply of REEs, secondary sources for REEs have been identified. These include phosphor-bearing cathode ray tubes and fluorescent lamps. It has been reported that about 15% of REEs produced are used in the production of fluorescent lamps and that fluorescent lamps contain 230 g of REE per ton. This content of REE is about 15 times higher than the content of REE in the primary ore resources. Waste phosphors from fluorescent lamps were sieved and the powder that passed 200 mesh size was used in one extraction. The chelating agent used was tri-n-butyl phosphate mixed with 70 wt% nitric acids. The phosphor sample was first activated by oscillation at 30 Hz for 60 min in a 25 ml tungsten grinding jar containing 15 mm diameter tungsten balls.

One g of the activated phosphor sample and 2 ml of the chelating agent were placed in a 100 ml high-pressure reactor and CO_2 was injected. The temperature and pressure were increased to 31°C and 73 bar, respectively. When the reaction was completed, the reactor was de-pressurized and Supercritical Fluid Extraction (SCFE) solution containing REEs was collected. An Inductively Coupled Plasma Optical

Emission Spectrometer (ICP-OES) was used to determine the Ce, Y, Eu, Tb, La, Sr and Sb contents in the SCFE solution at different wavelengths. The crystal morphology of the sample as received and after mechanical activation were also carried out.

21.6 Supercritical Fluids Applications

Supercritical Fluid Extraction (SCFE) has been reported as a green technology for metal extraction. The use of supercritical carbon dioxide fluid to extract REEs from primary and secondary resources has been reported. The technique involves lower energy consumption, higher recoveries of REE values and low waste generation. A supercritical fluid is one that has been heated such that it exists in a state above its critical temperature and pressure. The supercritical CO_2 fluid served as the solvent and tributyl-phosphate-nitric acid served as the chelating agent. The independent process variables that affect the extraction efficiency were identified as temperature, pressure, solid ore to chelating agent ratio, agitation speed and contact residence time. The fluid in the supercritical phase is neither a liquid nor a gas but it exhibits a combination of both properties such as the high diffusivity and low viscosity of gas and the solvation property of a liquid. The supercritical CO_2 is obtained by heating CO_2 above 31°C and at a pressure of 73 bar. Figure 21.3 shows the schematic setup of the supercritical fluid extraction apparatus.[9]

21.7 Recovery of REEs from Coal Resources

The treatment of coal-related materials for REE recoveries can be summarized as follows:

1. It has been reported that grinding to ultra-fine sizes is required to obtain a reasonable recovery during the physical processing of coal for REE recoveries. This is due to the complex mineralogy of the REE minerals and their very fine liberation sizes in coal.

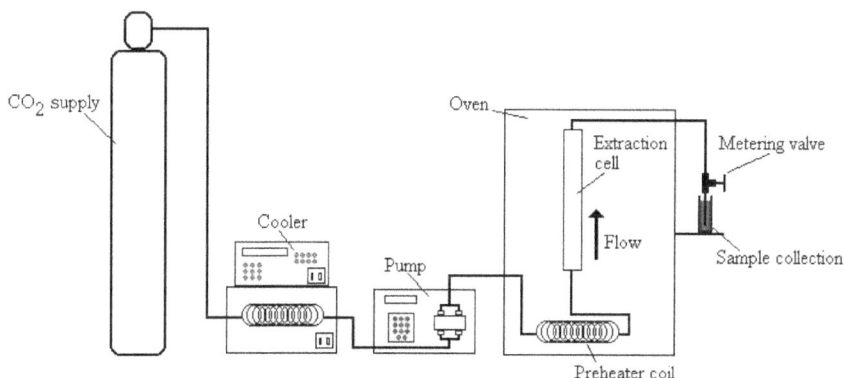

FIGURE 21.3 Schematic of SFE apparatus.

Source: Courtesy of Wikimedia (2022)

2. Direct acid leaching has been reported to produce moderate REE recoveries.

3. Apart from acid leaching, bio-leaching has also been applied in the recovery of REE from coal resources.

4. The application of leaching additives, and thermal and/or alkaline pre-treatment have been found to produce considerable improvements in the acid leaching process.

4. The leached solutions from coal have also been further purified by ion exchange and solvent extraction to improve the subsequent REE recoveries.

5. Pilot-scale studies have also been carried out on the leaching of coal-related resources to produce high-grade mixed REE at over 95% content.

The review of the various research studies carried out on coal and coal refuse to recover REEs showed that:

1. Gravity and magnetic-based separation techniques based on size, specific gravity and magnetic susceptibility factors have been found in-efficient with the maximum enrichment ratio of 1.21 obtained. However, the multiple stages of conventional and column froth flotation treatments of the decarbonized thickener underflow in a coal preparation plant using oleic acid as the collector produced concentrates that assayed 2300 and 4700 ppm, respectively. Oleic acid has also been used as the collector in the recovery of monazite and xenotime from heavy mineral sands. The main obstacle to the economic recovery of REE-bearing mineral particles from coals is that they are ultra-fine sized at less than 10 μm and are often interlocked in the host particle making excessive grinding to liberate the encapsulated particles necessary.

2. The leachability of REEs from coal using acid and salt leaching as well as leaching with pre-treatment has been investigated. The leaching of a high ash coal with ammonia sulphate led to the recovery of 80% REEs in the coal, but the purification of the leached solution with ion exchange was found to be inefficient. Furthermore, the leaching of lignite coal produced a recovery of about 90% using 0.5M sulphuric acid. The results obtained suggested that the leaching recovery of REEs depends mainly on the type of coal.

3. Alkali and thermal pre-treatments of coal and coal refuse have been applied to enhance the acid-leaching recoveries of REEs from coal and coal refuse. A decarbonized fine refuse was pre-treated with 8M sodium hydroxide at 75°C for two hours before it was acid leached. The results obtained showed that the recovery of REEs increased from 22% to 75% and HREE yield also increased from 38% to 48%. Alkali-acid leaching of coal refuse for simultaneous ash reduction and REEs recovery led to ash reduction from 46.21% to 14.17% and recoveries of 97% HREEs and 76% LREEs in the coal.

4. A coal gangue was subjected to roasting at 700°C for 30 min and then leached with 25% HCl leading to the recovery of 88.6% of the total REEs at room temperature. The activation by heat increased the crystallinity

of hematite, destroyed the inter-layered structures of clay minerals and improved leachability.

5. REEs were also successfully recovered from the fly ash residue of coal combustion which typically consists of 60–90% amorphous phase with the remainder as the. crystalline phase. Some advanced characterization techniques such as X-ray Absorption Near Edge Structure (XANES), micro-X-ray Absorption Near Edge Structure (μ-XANES), laser ablation inductively coupled plasma mass spectroscopy (LA-ICP-MS), multimodal image analysis, and sensitive high-resolution ion microprobe–reverse geometry SHRIMP-RG have been used to characterize the amorphous phase. When Sequential chemical extraction (SCE) tests were carried out on several classes of F-type fly ash samples that salsify the condition $SiO_2\% + Al_2O_3\% + Fe_2O_3\% > 70\%$, it was found that most of the REEs occurred in association with silicates and aluminosilicates with less than 10% REEs occurring in association with $CaCO_3$ and CaO. It was reported that the use of combined acid leaching and solution chemistry modeling resulted in 10–20% of REEs being leached in the pH range of 0–1.5, corresponding to the range where monazite and hematite dissolve based on solution chemistry modelling and hence 10–20% of the total REEs occurred in monazite and hematite forms.[10, 11]

References

1. Zhang, J., Watada, K., Sauber, M.E., and Azimi, G. (2020): *Supercritical Fluid Extraction of Rare-Earth Elements from a Canadian Ore.* TMS 2020 Proceedings, San Diego, CA.
2. Wikimedia (2022a): *Monazite* (https://upload.wikimedia.org/wikipedia/commons/c/cc/Monazite-%28Ce%29-164025.jpg, Accessed 28th May, 2022) (Attribution: Rob Lavinsky, iRocks.com—CC-BY-SA).
3. Britannica (2019): *Editors, Encyclopedia Britannica, Emulsion Chemistry* (https://www.britannica.com/science/emulsion-chemistry).
4. Wikimedia (2022b): *Bastanasite from Burundi* https://upload.wikimedia.org/wikipedia/commons/thumb/a/a2/Bastnaesit_Burundi.jpg/1024px-Bastnaesit_Burundi.jpg,160522 By Kouame—Own work, CC BY-SA 3.0 (https://commons.wikimedia.org/w/index.php?curid=8094492).
5. EURARE, Balomenos, E., Deady E., Yang, J., Panias, D., Friedrich, B., Binnemans, K., Seisenbaeva, G., Dittrich, C., Kalvig, P., and Paspaliaris, I. (2017): The EURARE Project: Development of Sustainable Exploitation Scheme for Europe's Rare Earth Ore Deposits (Final Report Summary). *Johnson Matthey Technology Review*, 61(2), 14 (https://doi.org/10.1595/205651317X695172).
6. Albert, C., Beladjine, M., Tsapis, N., Fattal, E., Agnely, F., and Huang, N. (2019): Pickering Emulsions: Preparation Processes, Key Parameters Governing Their Properties and Potential for Pharmaceutical Applications. *Journal of Controlled Release, Elsevier*, 309, 302–332. (https://doi.org/10.1016/j.jconrel.2019.07.003.hal-02330281).
7. Avazpour, R., Atifi, M., Chaouki, J., and Fradette, L. (2019): Physical Beneficiation of Rare Earth-Bearing Ores by Pickering Emulsification. *Mineral Engineering*, 144, 106034. (https://cordis.europa.eu/project/id/309373/reporting, Accessed 19th January, 2023).

8. Øistein Eriksen, D., Forrester, K.S., and Saxon, M.S. (2020): *Leaching of Eudialyte— The Silicic Acid Challenge.* TMS 2020 Proceedings (https://www.semanticscholar.org/ paper/Leaching-of-Eudialyte%E2%80%94The-Silicic-Acid-Challenge-Eriksen-Forres ter/637309e8a6a1eac217847183a4d26190f227c5e0).

9. Wikimedia (2022c): *Supercritical Fluid Extraction* (https://en.wikipedia.org/wiki/ Supercritical_fluid_extraction, Accessed 28th May, 2022) By Stainless316 at English Wikipedia—Transferred from en.wikipedia to Commons, Public Domain (https:// commons.wikimedia.org/w/index.php?curid=21379211; https://upload.wikimedia.org/ wikipedia/commons/c/c7/SFEschematic.jpg).

10. Honaker, R., Zhang, W., and Werner, J. (209): Acid Leaching of Rare Earth Elements from Coal and Coal Ash: Implications for Using Fluidized Bed Combustion to Assist in the Recovery of Critical Materials. *Energy and Fuels*, 83(7). (https://doi.org/ 10.1021/ acs.energyfuels.9b00295).

11. Wills, B.A., and Napier-Munn, T.J. (2006): *Mineral Processing Technology*, 7th Edition. Amsterdam: Elsevier Science and Technology Books.

22 Applications of Leaching and Microorganisms in Mineral Processing

22.1 Introduction

Chemical metallurgy methods of hydro and pyrometallurgy are used to alter the mineralogy of refractory ores to make them amenable to low-cost mineral processing methods. For example, hydrometallurgical atmospheric leaching was proposed for the recovery of metal value contents of the Australian McArthur River zinc-lead-silver deposit, while pyrometallurgical roasting is usually applied to convert the paramagnetic hematite to the ferromagnetic magnetite for more efficient magnetic separation. Some sulphide and oxidized copper ores have been pre-treated with leaching and contact reduction prior to froth flotation. The oxidized copper ore is dissolved in dilute sulphuric acid to recover the copper into solution and the copper pregnant solution is treated with iron scrap to recover the copper metal. Afterwards, the copper sulphide minerals that are not soluble are treated with froth flotation.

The method of chemical conditioning of a mineral surface such as sulphidization has also been used. In this case, the surface of oxidized ores is made to react with sodium sulphide and hence the surface becomes modified and it behaves as a pseudo sulphide with enhanced recovery response during froth flotation. The use of microorganisms in the leaching and froth flotation of ores has also been established. The bacteria thiobacillus ferroxidans have been used to enhance cyanide leaching of gold sulphide ores by breaking the sulphide lattice and thus freeing the gold particles that were occluded in them. It has also been shown that some bacteria will depress pyrite during the floating of coal.[1]

22.2 Application of Microorganisms

Microorganisms have been found to flocculate minerals that are finely dispersed. In bioleaching, metals from sulphide minerals are dissolved in acidic conditions using iron—and sulphur-oxidizing microorganisms. In addition, bio-oxidation in which gold-hosting minerals are pre-treated with microorganisms has been used in the mining industry. In bioflotation, microorganisms or their products are used as flotation collectors, depressants, and activators. It has been shown that the quantity of microorganisms for bioflotation needs to be high for bioflotation to compete favourably with conventional flotation chemicals. Bioflotation has been carried out with rhodococcus opacus in phosphate flotation and staphylococcus carnosus for fine coal tailings. and ferroplasma acidiphilum for pyrite. Furthermore, selective flotation of

DOI: 10.1201/9781003323433-22

sphalerite from a mineral mixture consisting of sphalerite and galena has been done with bacillus megaterium.

Halophilic microorganisms have been suggested for more environmentally friendly bioflotation processes in areas where seawater is abundant, and freshwater is scarce. It has been shown to produce various kinds of metabolites, such as poly-hydroxyalkanoates (PHA), ectoines, and bio-surfactants in addition to salt-tolerant enzymes. The adsorption of halophilic bacteria, specifically to pyrite, has been reported to occur within 5–10 min of exposure when the bacteria produce hydropho-bic moieties on their cell wall and thus act as pyrite bio-depressants.

When bioflotation microorganisms are adapted to minerals, the bioflotation per-formance may improve. For instance, when bacillus megaterium is adapted to the target minerals, better bioflotation performance was obtained. It is also impor-tant to note that microorganisms have also been reported to decrease the flota-tion process performance. Escherichia coli cells had been found to have a negative effect on the flotation efficiency of chalcopyrite, pyrite, and gold in a porphyry ore. Microbial cultures may thus have positive, neutral, or negative impacts on flota-tion. The metabolites and secreted compounds of microorganisms can also have an effect on flotation.

Bioflocculants are biopolymers, which form bridges in the agglomeration of fine particles. Bioflocculants have been analyzed and found to contain hydroxyl-, car-boxyl-, amino groups, polysaccharides and proteins. A biosurfactant produced by rhodococcus erythropolis was used as a flotation chemical for hematite and quartz. Microorganisms act on minerals by attaching to the surfaces, oxidizing, or reducing the mineral, and producing Extracellular Polymeric Substances (EPS) and polysac-charides. Additionally, the microbial activity can change the pH from 9.5 to 3 and Eh from +400 mV to −200 mV in a short time and can affect the chemicals used in flotation. Table 22.1 presents selected microorganisms for bioflotation.

Some bacteria, ferrobacillus ferooxidans and thiobacillus ferooxidans satisfy their energy requirements for living and growth from pure inorganic media by oxidizing

TABLE 22.1
Selected Microorganisms in Bioflotation of Ores

Microorganism	Bioflotation	pH range	Temperature (°C)
Leptospirillum ferrooxidans	Flocculation of chalcopyrite and pyrite. Greater depression of chalcopyrite compared	1.8–11	30
Acidithiobacillus ferrooxidans	Oxidation of pyrite surface and depressive effect on pyrite	1.8	30
Rhodococcus opacus	Use as biocollector and biofrother. Improved separation of apatite from quartz in phosphate flotation	2–12	20–28
Ferroplasma acidiphilum	Good depressant for pyrite	2.5–10	37
Bacillus megaterium.	Selective flotation of sphalerite from sphalerite-galena	2–10	

inorganic materials such as sulphur, ferrous iron and thiosulphate. The oxidation of pyrite and copper sulphide minerals are typical examples and occurs as follows.[2-4]

Pyrite, in the presence of moisture, undergoes a slow, non-bacterial oxidation as follows.

$$2FeS_2(s) + 7O_2(g) + 2H_2O(l) = 2FeSO_4(aq) + 2H_2SO_4(aq) \qquad (22.1)$$

When the reaction condition is made acidic or slightly alkaline, the iron (II) sulphate get oxidized fast and bacterial aided to iron (III) sulphate.

The iron (III) sulphate obtained gets hydrolyzed to yield sulphuric acid as follows:

$$Fe_2(SO_4)_3(aq) + 2H_2O(l) = 2Fe(OH)SO_4(aq) + H_2SO_4(aq) \qquad (22.2)$$

Furthermore, in the presence of oxygen and moisture, sulphur gets oxidized by bacteria to produce sulphuric acid as follows:

$$2S + 3O_2(g) + 2H_2O(l) = 2H_2SO_4(aq) \qquad (22.3)$$

Generally, reactions that produce acids provide conducive conditions for the growth of iron-oxidizing bacteria.

The iron (III) sulphate product of bacteria action can also attack sulphide minerals such as copper sulphide as follows:

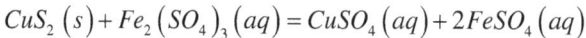

$$CuS_2(s) + Fe_2(SO_4)_3(aq) = CuSO_4(aq) + 2FeSO_4(aq)$$

References

1. SGS (2021): *Precious Metal Processing-Recovering Refractory Resources* (www.sgs.com, Accessed 15th October, 2021).
2. Kinnunen, P., Miettinen, H., and Bomberg, M. (2020): Review of Potential Microbial Effects on Flotation. *Minerals*, 10, 533.
3. Ghosh, A., and Ray, H.S. (1991): *Principles of Extractive Metallurgy*, 2nd Edition. New York: John Wiley & Sons.
4. Wills, B.A., and Napier-Munn, T.J. (2006): *Mineral Processing Technology*, 7th Edition. Amsterdam: Elsevier Science and Technology Books.

23 Tailings Re-treatment

23.1 Introduction

The material remaining after the recovery of concentrates from an ore feed, called tailings, has for a long time been a burden to the mining industry in terms of the cost of its management, disposal difficulty and liability risks. Tailings which were traditionally considered waste are now been considered for retreatment due to decreasing ore grades, improved ore processing technology and a new drive for sustainable resource exploitation. They are now regarded as un-mined treasures. The large mass outputs of tailings make its management a big concern, while its toxic nature poses a risk to humans and the environment thus creating a concern regarding its safe disposal. The methods to process tailings depend on the type of valuable mineral, the methods of beneficiation applied and the particle size of the tailings, among others.

Tailings can be classed as hand picking, gravity, magnetic separation, flotation, chemical beneficiation and electrification tailings. Tailing slurries are usually deposited in ponds where they are isolated by laying a cover on them or treated with chemicals to eliminate their toxicity or both. They are also used in backfilling of mined pits.

The run-of-mine ores bearing mineral values hosting metals such as gold, copper and uranium are taken through several processing stages like comminution, physical and chemical based separations, leaching and leached solution purifications such that the tailings bearing most of the gangues become contaminated with chemical reagents added at various stages. Tailings are in the form of a slurry and are kept in Tailings Storage Facilities (TSF) such as tailings ponds and dams. The failure of tailings storage facilities has been reported. For instance, the failure in 1998 of Spain's Los Fraile zinc mine tailings dam led to the spilling of more than one billion gallons or about 3.79 billion litres of acidic, heavy metal-laden sludge into the surroundings and into a river called Guadiamar.[1]

The spill decimated the surroundings by contaminating soil, water and vegetation and caused a mass extinction of wildlife with other effects on the ecosystem that could not be reversed till date. It has also been pointed out that even stable tailings dams can pose problems such as dust contamination from dry tailings, release of radioactive elements, leaching of the adjacent land mass and acid mine drainage. The reuse or re-claiming of tailings is being driven not only by the need to reduce the risk of its storage but its ready availability and rapid depletion of richer ore deposits as well as improved extraction technology that makes tailings increasingly attractive. The toxic chemicals, wastewater and heavy metal ions in mineral processing will not only pollute the environment and cause ecological damage, but will also increase the potential safety hazard of the tailings dam and may induce landslides, mudslides, tailings dam breaks, and other accidents thus making tailings treatment a highly valued option.

DOI: 10.1201/9781003323433-23

23.2   Approaches for Tailings Re-treatment

23.2.1   Industry Re-treatment of Tailings

Several mining industry players have initiated efforts to retreat tailings. For instance, Sibanye Stillwaters estimated about 175.766.85 tons of gold and 44.13 million tons of uranium in its tailings storage in the area called the Golden arc stretching from Johannesburg to Welkom in South Africa. Pan African Resources also plans to treat one million tons of tailings per month and is estimated to recover 19.5 tons of gold over the project's lifetime. Tailings are also being considered for use as construction materials such as in concrete and mortar making and the like. Such applications are expected not only to relieve the burden of managing large quantities of tailings but would also reduce the mining of natural aggregates which will require substantial energy. Ore tailings are also retreated by froth flotation and smelting to recover the low values in them. The tailings retreatment methods include the following[1]:

23.2.2   Retreatment through Heap Leaching

The retreatment of tailings varies depending on the source of tailings and the material(s) to be recovered. The heap leaching and Solvent Extraction—Electrowinning processes have been the key methods in treating tailings. In heap leaching, the ore tailings are agglomerated and stacked on a heap and then irrigated with a leached solution and it is an ideal method for low-grade ores. The benefits of heap leaching include lower operating costs, lower capital costs and simple design and operation among others. The combined solvent extraction—electrowinning process route enables a highly effective, cost-efficient extraction and separation of the target material from low-grade ores.

23.2.3   Tailings Re-concentration

It is known that the contents of the mineral values in tailings are relatively low in comparison to the parent feed. Therefore, advanced processing technologies such as froth flotation can be used to recover additional mineral values economically. For instance, flotation with new chemical reagents and enhanced techniques has been used to re-concentrate tailings bearing mineral values such as apatite and zinc oxide. New gravity separation equipment such as vertical centrifugal concentrators, composite gold shakers and compound centrifugal concentrators can recover values from fine-grained and ultra-fine tailings.[2]

23.2.4   Microbial Treatment and Leaching Technology

In this technology, the biochemical action of microorganisms is used to improve the tailings' properties and remove harmful metal ions in tailings through microbial leaching or microbial mineralization to avoid environmental pollution.[3]

23.2.5   Tailings Dewatering Process

The tailings from the mineral beneficiation stage usually carry high water content. The high water content in the tailings may result in a tailing dam accident. Therefore, there is a need de-hydrate or dewater the tailings. Tailings are dewatered using hydrocyclone, filter press and by the use of dewatering type cyclone, high-frequency dewatering screen and deep cone thickener to obtain tailings containing less than 15% moisture which is called a dry discharge. The filter press also produces filter cake with low moisture content while the filtrate water is recovered for re-use. The FLSmidth tailings dewatering system is used for coal tailings. It is enhanced with a feature called 'S rolls' in some important areas to double their filtration areas and so improve the cake's filtration efficiency.[4]

23.2.6   Filling of the Mined-out Areas

The tailings are used as waste to refill the mined area for the protection of the terrain and to reduce the cost of the tailings dam. The refilling may be of the following two types.

23.3   Tailings Sand Cement Filling Technology

The tailings are mixed with cement, ash, and others to form a filling aggregate with a hardness of 1–2 MPa, and then fed into the mined zone by pipeline self-flow.

23.4   Water Consolidation Sand Filling Technology

High-water consolidation full-tail sand filling is similar to full-tail sand cementing, with the exception that high-water content materials are used as cementitious materials rather than cement. This method uses alumina, sulpho-aluminate-based material and agglomerate, quicklime, and coagulant-based material, which are separately made into pulp and mixed prior to filling. They exhibit a fast speed of setting.

23.4.1   Production of Geo-polymers

A new research carried out in Finland aims at creating wealth from the waste tailings by making them into geopolymers that are similar to concrete or ceramics because of their typical high contents of aluminosilicate. The alkali-activated tailings are subjected to mechanical pre-treating to increase their compressive strength and further heat treatment to convert them to high-strength ceramics.[5, 6]

References

1. Ebbenis, A., and Carlos, C. (2021): *Tailings Retreatment: The Next Big Revenue Source for the Mining Industry* (www.feeco.com, Accessed 15th October, 2021).
2. JXSC (2019): *Mine Tailings Retreatment Process-JXSC Machines* (https://www. jxscmachine.com/new/mine-tailings-retreatment-process/, Accessed 15th October, 2021).

3. Kinnunen, P., Miettinen, H., and Bomberg, M. (2020): Review of Potential Microbial Effects on Flotation. *Minerals*, 10, 533.

4. Smidth, F.L. FLSmidth Upgrades Tailings Dewatering Presses. *Mining Magazine* (Accessed 21st March, 2022).

5. University of Oulu (2021): *Towards Safe and Sustainable—The Geomins Project Provides Alternative Methods to Handle Mine Tailings.* Oulu: University of Oulu. (https:////www.oulu.fi/en/news/towards-safe-and., Accessed 19th January, 2023).

6. Wills, B.A., and Napier-Munn, T.J. (2006): *Mineral Processing Technology*, 7th Edition. Amsterdam: Elsevier Science and Technology Books.

24 Mineral Processing of Beach Sand Minerals

24.1 Introduction

Sand can be described as loose grains that cover the beaches, deserts and river beds of the world. The earth's landmass is made up of rocks consisting of minerals like quartz, feldspar and mica. The weathering action of wind, rain and ice disintegrates the earth's rocks and minerals into grains of smaller sizes. However, quartz is more resistant to weathering because it is hard and water-insoluble in comparison to other rock minerals and is thus largely retained. The quartz grains are transported by wind, rivers, streams and waves to the seashore where they get accumulated as beach sands of different colors. Beach sand may be of three types, that is, white, light-colored or tropical white sand. The white sand consists mainly of quartz and limestone, while light-colored sand derives its color from its quartz and iron contents. On the other hand, the tropical white sand or sands of the tropical islands are typically white because they contain calcium carbonate from the remains of shells and skeletons of reef-dwelling organisms like corals. The sands of tropical beaches may also appear black when they contain glasses of volcanic origin. Furthermore, tropical beach sand can also have green color when forces of erosion get olivine separated from other associated minerals.

The beach sand of Egypt, also called black sand, contains several minerals of economic value such as ilmenite and monazite. The reserves of economic minerals in the top 20 m of Egyptian beach sands have been estimated to be over 616 million metric tons and they are found either as beach sand sediments or coastal sand dunes. The monazite grains exhibit colors such as lemon yellow, and light to deep canary and the content can be up to 0.6% by weight in thin layers of 10–30 cm near and parallel to the shoreline. The beach sand monazite hosts high contents of cerium, lanthanum and neodymium and is known to be particularly rich in the lighter rare earth elements, the cerium sub-group. The major part of the world's supply of titanium-bearing minerals is in beach sands such as those of South Africa and Sri Lanka. Sri Lanka's North East coast has many locations endowed with naturally occurring concentrates of ilmenite and rutile.[1,2]

The North East coast of Sri Lanka hosts main deposits that contain 80% heavy minerals and assay 70–72% ilmenite and 8% rutile. In a research study, one hundred tons of beach sand covering a stretch of 2 km was collected. The sands were scraped manually from the surface down to depths of between 10 and 30 cm. The ore was beneficiated using a combination of wet gravity, low- and high-intensity magnetic separation as well as electrostatic separation to obtain concentrates that assay 97% monazite by weight and at a recovery of about 77%. The beach sand monazite is typically recovered by first subjecting the beach sand to shaking tabling. The tabling

DOI: 10.1201/9781003323433-24

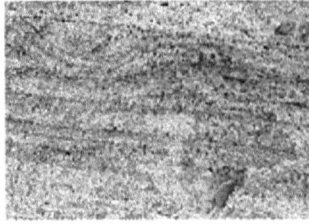

FIGURE 24.1   Heavy minerals (dark) in Chennai quartz beach sand of India.

Source: **Courtesy of Wikimedia (2022)**

concentrate is then treated with magnetic and electrostatic separators to recover monazite concentrates that need to be further treated to obtain market specifications. The electrostatic separator is used to separate monazite from zircon and other associated minerals by making adjustments such as voltage, the radial distance between the ionic electrodes and carrier rotor surface, ore feed rate and the angular position of the static electrode. The monazite concentrate is efficiently dissolved in 2M sulphuric acid and thorium and uranium are separated from each other by cation exchange.[3]

The presence of monazite and xenotime grains were confirmed with an electron microprobe analyzer (EMPA), while the rare earth elements (REEs) contents were determined by comparing the X-ray counts obtained with that of a synthetic glass doped with REEs. The entire polished surface was examined under backscattered electrons at 15 kV, 30 or 50 nA electron current, a counting time of 750 s and with the electron beam opening diameter of 3 or 6 μm based on the grain size. For the analysis of light elements such as calcium, phosphorous, silicon and iron the Kα lines were used while the Lα lines were used for most of the REEs. Figure 24.1 shows an Indian beach sand physical feature indicating the presence of heavy minerals in quartz sand.[4]

24.2   Processing of Beach Ores

High-intensity magnetic separators, induced roll and disc magnetic separators are widely used to treat beach sands. For instance, the North East coast Pulmoddai plant used a wet magnetic separator to recover ilmenite from beach sand. However, some processing plants use dry magnetic separation to recover ilmenite and leucoxene from mineral concentrates of high density. For the Sri Lanka beach sands, Tel and Sabah (2016)[5] reported that mineralogical and chemical analyses indicated that 71% of beach sand particles were smaller than 355 μm with greater than 99% of the titanium content in this fraction. The sand was processed as follows:

1. The size fraction that passed the 355 μm sieve but retained on the 63 μm sieve was used
2. The sample was treated in a dense medium with TBE as the medium to produce light and heavy fractions and the latter was processed further with

a low-intensity magnetic separator to obtain a magnetic concentrate, defined as concentrate 1 and non-magnetic minerals delivered to Cook isodynamic magnetic separator (CIMS) to remove magnetic contaminants

3. The magnetic contaminants from CIMS were further treated in magnetic fields with strength 0.1, 0.3 and 0.5 T to produce concentrates 2, 3 and 4, respectively

The beach sand of Tirikaye contains magnetite with 48.41% iron and was treated with a dry magnetic separator to obtain concentrates. For instance, when the beach sand fraction was treated by a dry magnetic separator at about 0.0075 T magnetic fields, a high grade concentrate was obtained.

References

1. LiveScience (2021): *What Is Sand?* (www.livescience.com/34748-what-is-sand-beach-sand.html, Accessed 21st March, 2022).
2. Moustafa, M.A., and Abdelfattah, N.A. (2010): Physical and Chemical Beneficiation of Egyptian Beach Sand Monazite. *Resource Geology*, 60(3), 288–299.
3. Premaratne, W.A.P.J., and Rowson, N.A. (2004): The Processing of Beach Sand from Sri Lanka for the Recovery of Titanium Using Magnetic Separation. *Physical Separation in Science and Engineering*, 12(1), 13–22.
4. Wikimedia (2022): *Heavy Minerals (dark) in a Quartz Beach Sand (Chennai, India)* (https://commons.wikimedia.org/w/index.php?curid=4175108 By Photograph taken by Mark A. Wilson (Department of Geology, The College of Wooster) [1]—Original photograph, Public Domain, Accessed 22nd April, 2022).
5. Tel, M., and Sabah, E. (2016): Beneficiation of Beach Magnetite Sand, Pamukkale University. *Journal of Engineering Sciences*, 22(3), 220–225.

25 Processing of Battery Minerals

25.1 Introduction

The transition by nations to a low carbon economy in a bid to halt global warming and climate change has led to a shift of attention to the rare earth metals and other critical minerals like those of lithium, cobalt, nickel, graphite, tantalum and vanadium called battery minerals, to support their climate commitment and economies. Battery minerals are mineral-bearing metals which are used in rechargeable batteries. From a historical account, it is known that the driving force for lithium and cobalt search has been the demand for batteries and primarily for consumer electronics.

The battery minerals, particularly lithium and cobalt are experiencing supply challenges because apart from their need in the energy transition to combat climate change, they are also increasingly used in electric vehicles, e-bikes and electrification of tools as well as in battery-based energy storage systems. The European Union has postulated that by 2050, the demand for lithium would be 280 per cent of the current known deposits while that of cobalt will be 426 per cent of the same. China is presently the lead producer either by the owner of these critical minerals or by investment, or global ownership of their mines coupled with local processing capability. There is currently a hot competition between China, the United States and European Union states to have access to these strategic minerals and recycling of battery minerals as in line with the circular economy is part of the efforts being made. In March 2001, a Singapore-based e-waste recycler started operation and was able to recycle 14 tons of lithium-ion batteries per day and successfully recover 90% of its precious metals such as lithium and cobalt.

Western Australia is richly endowed with hard rock lithium deposits and is the lead world producer of lithium minerals as it accounts for 41% of the world's supply. It is also the second-world producer of cobalt minerals and the leader in rare earth minerals at 16% of global production Lithium has a density lower than that of water and it is soft and possesses excellent energy density so that it can store more energy in comparison to its size. Between 2017 and 2018, Western Australia produced 2.1 million tons of spodumene, the lithium mineral estimated to be worth $1.6 billion. About forty Australian gold and nickel mines are found to contain cobalt too. Democratic Republic of Congo (DRC) and Australia are notable world cobalt producers.[1,2]

DOI: 10.1201/9781003323433-25

25.2 Beneficiation Methods

25.2.1 Beneficiation of Nickel Ores

Otunniyi et al. (2016)[3] reported that the Botswana Selebi-Phikwe complex ore of nickel, copper and cobalt was concentrated by froth flotation with copper activation treatment to recover copper and nickel. The ore run-of-mine assayed 0.55–0.71% Ni and 0.58–0.75% Cu. For the ore, pentlandite was found finely disseminated in pyrrhotite making its liberation difficult. The ore was treated with different dosages of copper sulphate as the copper activator and potassium normal butyl xanthate as the collector. The Ni recovery was observed to increase with increasing copper sulphate dosage up to a maximum and it then decreased. The cumulative recovery of Co was found to increase to the highest of greater than 60% at the highest copper sulphate activation dosage.

The oxidized copper and cobalt ores are first subjected to sulphidization and then processed by froth flotation to reduce the gangue contents. Some of the oxide gangue minerals were converted to sulphides which exhibit the least hydrophilicity of all polar minerals and are thus rendered more amenable to flotation. Figure 25.1 presents the flowsheets for the processing of nickel-bearing oxide and sulphide ores.[4]

25.3 Leaching Recovery of Cobalt and Nickel

Nickel and cobalt are usually found coexisting in their ores such as nickel-copper sulphides and nickel laterite ores as well as in secondary sources like used nickel-metal hydride lithium-ion batteries. The separation of cobalt from nickel is however difficult because the two metals exhibit similar chemical properties. Generally, the cobalt and nickel recoveries from the primary and secondary sources will involve ore/waste preparation/comminution, selective leaching, purification of leach solution and electrochemical recovery of nickel and cobalt. Leaching with sulphuric acid is preferred to hydrochloric acid because the former is cheaper and less corrosive.

The split anion solvent extraction technique that uses chloride ionic liquids can be applied in separating Co(II) in sulphate media from Ni(II). This is due to the fact that the two are present in anionic sulphate forms SO_4^{2-} and HSO_4^{2-} which are more hydrated than the chlorides in the ionic liquids in addition, the tendency of Co(II) to form tetrachlorocobaltate (II) in chloride ionic liquids is greater than that of Ni(II).

The hydrometallurgical processing of a Turkish nickel-cobalt complex ore was reported by Kursunoglu and Kaya (2019)[5]. Solvent extraction (SX) was applied to separate and purify nickel and cobalt in a sulphate leach solution obtained by leaching a laterite nickel ore.

25.4 Processing of a Nickel-copper Sulphide Ore

911 Metallurgist (2022)[6] reported the processing of a complex copper-nickel-cobalt ore to recover nickel and copper as follows.

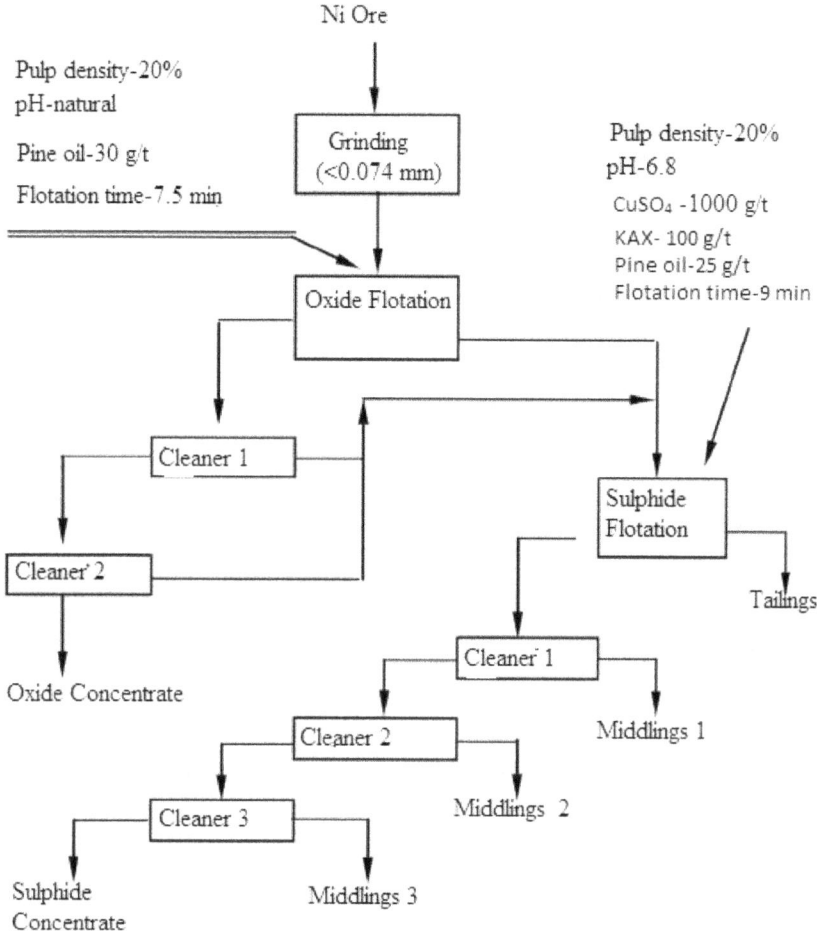

FIGURE 25.1 Beneficiation flowsheets for the beneficiation of oxide and sulphide nickel ores.

Source: Yuce et al. (2008) (courtesy of Taylor and Francis, 2022)

25.4.1 Flotation Concentration

The magmatic sulphide ore is hosted in an altered basic rock and the drill core assayed 0.8%, 0.3%, 0.6%, 5.2%, 36.5% and 13.3% of nickel, cobalt, copper, sulphur, silica and iron, respectively. The minerals in the ore consist of pyrite, pyrrhotite, pentlandite and chalcopyrite. The pentlandite often occurs as fine intergrowths within the pyrrhotite with a liberation size in the range of 10–20 μm. The ore passing 10 mesh sieve was ball milled and floated by first conditioning with soda ash and amyl xanthate at 0.68 and 0.18 kg/ton dosages for 5 min, respectively. The duration of the flotation was about 20 min.

25.4.2   Roasting

Dried and finely ground sulphide concentrate filtrate cake mixed with between 7% and 8%, sodium sulphate was subjected to a sulphatizing roast by charging through a screw feeder into a 10.2 cm diameter and 22.9 cm long fluidized bed roaster swept at the end by the main fluidizing air which entered the bottom of the bed. The bed depth was kept at 9.14 cm but could be altered with a dry feed rate of 24 g/min implying a retention time of about six hours. The roasting temperature ranged between 670 and 690°C and the calcine bed product obtained contained about 86% nickel.

25.4.3   Leaching and Solution Recovery

The roast calcine was leached in water and the optimum result was obtained at 80°C and a one-hour leaching duration. The iron content was removed using pure calcium carbonate in a neutralization reaction at a temperature of 70°C to produce a cake.

25.4.4   Copper Solvent Extraction

Copper and nickel were separated by solvent extraction with naphthenic acid in kerosene after partial neutralization of the iron-free liquors with ammonium hydroxide. Zinc and residual copper impurities were removed with hydrogen sulphide, while cobalt impurity was removed with chlorine. Copper was then separated from nickel because there is an appreciable difference in the pH at which the two metals are extracted. Copper can be extracted, with very small amounts of nickel contamination at pH 6.0–6.5, while nickel is extracted from copper at pH 8.0–8.2. It was shown that 86 per cent of the nickel and 95 per cent of the copper were leached from the calcine.[7, 8]

25.4.5   Processing of Lithium Ores

Lithium is an alkali metal with a color that varies from silver- white to grey. It is a very reactive metal that has the least density (534 kg/m^3) of all non- gaseous elements at a temperature of 20°C. The lithium bearing minerals found in nature as economic deposits include spodumene ($LiAlSi_2O_6$), found in pegmatite, petalite ($LiAlSi_4O_{10}$) and jadarite ($LiNaSiB_3O_7(OH)$). Lithium is found dissolved in continental brine deposits and in the salty sea water at a concentration of 0.18 ppm in a total dissolved solids of 35000 ppm. The Manono pegmatite of Democratic Republic of Congo has been reported to host over 2 million tons of Li metal.

Li minerals found in hard rock are excavated by surface or sub-surface mining, while those in brines are obtained by pumping from wells.

The Li bearing minerals are typically concentrated using the following methods:

1. The ore is subjected to further crushing as required by the separation techniqueto be used
2. The crushed ore is then gravity concentrated to remove the denser gangue minerals such as cassiterite and tantalite

3. Froth flotation is then applied to the gravity concentrate to remove quartz and feldspar
4. Acid washing is applied to the flotation concentrate to remove apatite
5. Wet High intensity magnetic separation can finally be applied to the acid washed concentrate to remove paramagnetic minerals like tourmaline

The final concentrate is subjected to acid or carbonate leaching to produce lithium carbonate.

The carbonate leaching is carried out under pressure (2140 kPa) at a temperature of 215°C to produce lithium carbonate which is converted to the more soluble lithium bicarbonate. The condition of the pregnant solution of lithium bicarbonate is adjusted to precipitate impurities like sodium and iron. The purified solution is further heated to precipitate lithium carbonate.[9]

References

1. Business for Social Responsibility (BSR). *A Scramble for Battery Minerals: Emerging Issues.* BSR (https://www.bsr.org/en/emerging-issues/a-scramble-for-battery-minerals).
2. Chamber of Minerals and Energy, Australia (CME) (2021): *Lithium & Battery Minerals—The Chamber of Minerals and Energy of Western Australia* (www.cmewa.com.au).
3. Otunniyi, I.O., Oabile, M., Adeleke, A.A., and Mendonidis, P. (2016): Copper Activation Option for a Pentlandite–Pyrrhotite–Chalcopyrite Ore Flotation with Nickel Interest. *International Journal of Industrial Chemistry,* 7, May.
4. Yuce, A.E., Bulut, G., Boylu, F., and Onal, G. (2008): Flowsheet Development for Beneficiation of Nickel Ore. *Mineral Processing and Extractive Metallurgy Review,* 29(1), 57–67.
5. Kursunoglu, S., and Kaya, M. (2019): Hydrometallurgical Processing of Nickel Laterites-a Brief Overview on the Use of Solvent Extraction and Nickel/Cobalt Project for the Separation and Purification of Nickel and Cobalt. *Scientific Mining Journal,* 88, March.
6. 911 Metallurgist (2022): *Copper-Nickel Ore Processing* (https://z4y6y3m2.rocketcdn.me/wp-content/uploads/2019/11/Copper-Nickel-Ore-Processing-Preliminary-Flowsheet.jpg, Accessed 22nd April, 2022).
7. Onghena, B., Valgaeren, S., Hoogerstraete, T.V., and Binnemans, K. (2017): Cobalt(II)/Nickel(II) Separation from Sulfate Media by Solvent Extraction with an Undiluted Quaternary Phosphonium Ionic Liquid. *Royal Society of Chemistry Advances,* 7, 35992.
8. McKinsey & Company (2021): *Lithium and Cobalt: A Tale of Two Commodities* (https://www.mckinsey.com/industries/metals-and-mining/our-insights/lithium-and-cobalt-a-tale-of-two-commodities, Accessed 5th May, 2021).
9. British Geological Survey (BGS) (2016): Lithium (https://www2.bgs.ac.uk/mineralsuk/download/mineralProfiles/lithium_profile.pdf, Accessed 21st March, 2023).

26 Processing of Refractory Ores

26.1 Introduction

Refractory ores are ores that are considered difficult to process for the recovery of concentrates because either they have mineralogy which is complex or the mineral values are finely disseminated or both and hence the conventional mineral processing methods perform poorly in treating the ores.

The large Australian McArthur River zinc-lead-silver deposit that assayed 13% Zn and 6% Pb in 2003 and could not be economically concentrated for over three decades because of its fine-grained texture is an example. The ore only became possible to process when the IsaMill fine grinding technology was introduced and used with a suitable flotation technology when the mine was opened in 1995. The lead-zinc concentrates produced were 80% < 7 μm ultra-fine size consist. It is also common knowledge in the mining industry that directly leachable gold deposits are getting depleted and that for the foreseeable future there will be an increase in the proportion of gold extracted from refractory ore deposits.[1] It has been reported that from 1994 to 2009 the percentage of global gold extraction from refractory gold ores increased from 5% to over 10%. Gold ores can be classified as follows:

1. Free milling ores that yield over 95% gold recovery in solution during standard cyanidation treatment
2. Refractory gold ores that can be subdivided as mildly, moderately and highly refractory such that the gold yields into solution under standard cyanidation treatment are 80–95%, 50–80% and less than 50%, respectively

Refractory ores generally consist of sulphide minerals, organic carbon, or both. The sulphide minerals are known to be impermeable to the cyanide leaching solution because they occlude the ultra-fine gold particles and thus make it difficult for the leach solution to form a water-soluble stable complex with gold. On the other hand, organic carbon present in gold ore may adsorb dissolved gold-cyanide complexes in a way similar to the activated carbon and thus robbing the activated carbon of part of the dissolved gold.[2]

Refractoriness of gold-bearing ores arises because the ultra-fine gold particles are encapsulated in a host mineral, usually, pyrite or arsenopyrite that is impervious to cyanide solution and thus making the gold value inaccessible to the leaching cyanide solution. Refractoriness is a measure of the degree of resistance of an ore to the leaching of its gold content by standard recovery methods. In view of this, refractory ores have to be pre-treated, either by some physical or chemical methods to be able to make adequate value metals recovery achievable by the traditional route of direct

DOI: 10.1201/9781003323433-26

cyanide leaching and carbon adsorption. An ore refractoriness, that is, its degree of resistance to standard recovery methods is generally due to the extremely fine gold particles being fully encapsulated by a host mineral that the cyanide leach solution finds impenetrable.[2]

26.2   Pre-treatment Methods

The pre-treatment methods include the physical roasting and ultra-fine grinding to liberate the encapsulated gold particles as well as the hydrometallurgy-based pressure oxidation of the ore as received as well as pressure and tank bio-oxidation of concentrates of the flotation process. The ore lumps are first subjected to crushing, grinding, gravity separation and froth flotation before pre-treatment. Roasting is usually applied to sulphide and refractory ores and it is conducted in the temperature range of between 500°C and 700°C. The daily throughput per line can be as high as 5,000 tons.

During roasting, the sulphur in the mineral is converted to sulphur dioxide, while pressure oxidation and bio-oxidation will produce sulphate anions from the sulphur into the solution. When iron (II) sulphate and jarosite ($KFe_3(SO_4)_2(OH)_6$) are subjected to pressure oxidation and bio-oxidation, they will yield soluble iron (III) sulphate. After pre-treatment, the ore is taken through intense cyanidation. The roasting reactor can be of bubbling fluidized bed (FB) or circulating fluidized bed (CFB) type and the oxidizing agent can be pure oxygen or air. The roasting process can be used to remove impurities such as arsenic and mercury and to recover sulphuric acid as well as heat energy from the process. For environmental safety, sulphur dioxide and arsenic trioxide must be efficiently removed from the process gas.

It is a common practice to make some sulphide ores water or acid soluble by oxidizing or chloridizing roasting. The oxidizing roastings of sphalerite that produce acid-soluble oxide and sulphate take place as shown in the following equations.[3]

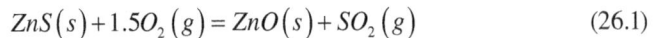

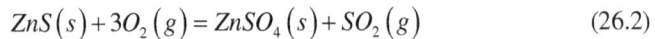

$$ZnS(s) + 1.5O_2(g) = ZnO(s) + SO_2(g) \tag{26.1}$$

$$ZnS(s) + 3O_2(g) = ZnSO_4(s) + SO_2(g) \tag{26.2}$$

The chloridizing roasting of magnesium oxide takes place as follows:

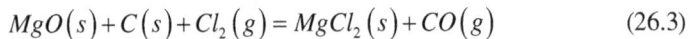

$$MgO(s) + C(s) + Cl_2(g) = MgCl_2(s) + CO(g) \tag{26.3}$$

The magnesium chloride produced is dissolved in alkali chloride at 700°C and electrolyzed to recover magnesium. Pressure oxidation, or autoclaving, was initially developed to process base metal concentrates. In the last three decades, it has been used to process refractory gold ores and concentrates having gold particles trapped in sulphide minerals, such as pyrite or arsenopyrite. During the process, the gold-bearing sulphide minerals in an aqueous slurry form are oxidized by pure oxygen at a high temperature and pressure. This treatment converts the sulphides to stable and water-soluble phases consisting of metal sulphate compounds and sulphuric acid. The gold locked in the original sulphide mineral is completely liberated, allowing very high gold recovery to be achieved when the product is treated with cyanidation. The use of pressure oxidation has become common in the past two decades because of its better efficiency in gold recovery in comparison to roasting. In addition, the first-generation of roasting plants have not been able to meet increasingly strict environmental controls on sulphur and arsenic discharge.

However, pressure oxidation has been found unsuitable to treat ore feeds with high content of silver because silver often reacts with iron in the autoclave to form a silver jarosite compound that resists leaching by cyanide. The liberation and recovery of the silver require expensive techniques. It has also been known that during pressure oxidation, only a small proportion of the iron in pyrite and arsenopyrite is converted to hematite and the major part ends up as jarosite or basic iron sulphate. These compounds have the potential to pollute the environment with heavy metals and can adversely affect downstream processes to recover precious metals.[4]

In bio-oxidation, certain strains of bacteria are used to accelerate the natural process of sulphide oxidation. These bacteria obtain energy from the oxidation of ferrous ions and sulphides. The by-products then report to the aqueous phase as metal sulphate compounds and sulphuric acid. Gold recovery in the process is typically very high and compares favorably to what is obtainable in pressure oxidation.

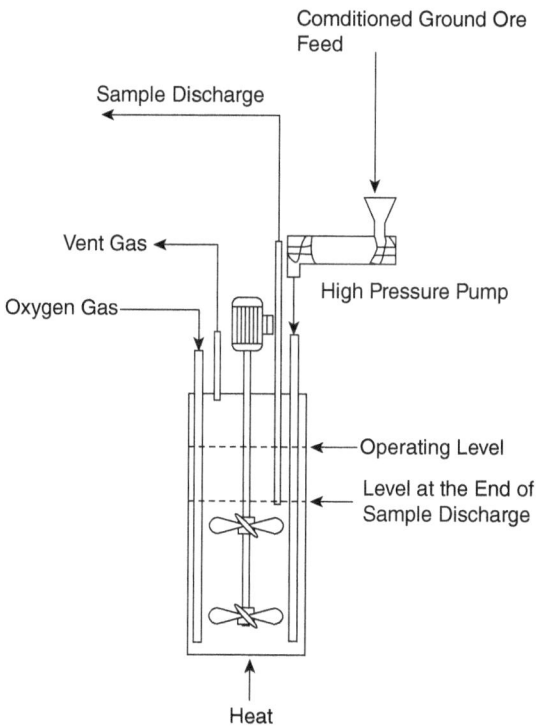

FIGURE 26.1   A typical semi-continuous autoclave for pressure oxidation.

Source: **Courtesy of 911 Metallurgist (2022)**

Bio-oxidation is a reasonably well-established technology with several plants operating globally. In bio-oxidation, high-grade concentrates are finely milled in stirred tank reactors before treatment. The bacteria are autocatalytic, that is, the bacteria which serve as the catalyst for the reaction also benefit from it through multiplication.

An autoclave is used for pressure oxidation of a slurry material comprising at least one sulphide material. The autoclave is made up of a pressure vessel that holds the slurry material to be treated. The pressure vessel consists of units that are arranged in horizontal positions one after the other and are separated by one or more partitions. The partition is equipped with an upper edge or at least one opening to define the pulp level of the unit. It also has an inlet that receives oxygen-containing gas into the pressure vessel. The autoclave is equipped with a stirrer having an upper impeller and a lower impeller mounted in a vertical shaft. Figure 26.1 presents the schematic diagram of an autoclave.[5–7]

References

1. Wills, B.A., and Napier-Munn, T.J. (2006): *Mineral Processing Technology*, 7th Edition. Amsterdam: Elsevier Science and Technology Books.
2. SGS (2021): *Precious Metal Processing-Recovering Refractory Resources* (www.sgs.com, Accessed 15th October, 2021).
3. Ghosh, A., and Ray, H.S. (1991): *Principles of Extractive Metallurgy*, 2nd Edition. New York: John Wiley & Sons.
4. Patent CN112657423A China (CN112657423A—Autoclave and Pressure Oxidation Process—Google Patents, 13th November, 2021).
5. 911 Metallurgist (Pressure-Oxidation-Semicontinuous-Autoclave-Apparatus-1.jpg (530 × 533) (www.911metallurgist.com, Accessed 13th November, 2021).
6. John T. Boyd Company. *John T. Boyd Company—Refractory Ore Specialist Consultant—Gold and Metals Services Fulfilling the Needs to a Diverse International Metals Mining Client Base* (www.jtboyd.com, Accessed 13th November, 2021).
7. Kinnunen, P., Miettinen, H., and Bomberg, M. (2020): Review of Potential Microbial Effects on Flotation. *Minerals*, 10, 533.

27 Processing of Non-metallic Ores

27.1 Introduction

Resource refers to a naturally occurring commodity of economic value found in an area, while the reserves of such a resource refer to the amount of the resource that can be exploited at a profit with the existing technological know-how and those that are presently unsuitable for extraction either for economic, technological or political reasons. The resources of the earth can also be categorized as renewable and non-renewable. The renewable resources are those that get re-produced or replenished within a short time span, while non-renewable resources are those that took a very long time that can be millions of years to form. Renewable resources include direct solar energy and energy obtainable from flowing rivers and wind, while non-renewable resources are exhaustible like metallic and non-metallic materials whose reserves quantities are fixed.

A mineral is a naturally occurring solid, inorganic material found in the earth's crust and possesses a definite chemical composition and a regular crystalline structure. A rock is a solid material comprising of one or more minerals. An ore or an ore body is a natural aggregation of different minerals resulting from mineralization processes, with some of the minerals being high in concentrations of metals and most of the time it is enclosed in a country rock considered barren or a waste. On the other hand, an ore deposit is an accumulation of minerals or metals with sufficiently high concentrations that it can be processed to extract its value at a profit with the current technology and under the prevailing economic conditions. Such a deposit is specifically called a commercial deposit. An ore may be metallic such as galena or non-metallic such as fluorite and sulphur and to be considered of value, its elemental compositions must be above their average crustal abundance. Ores can be classed as high-grade low-value metals and low-grade high-value metal ores. This means that the ore may be of high grade in terms of the contained metal which is of low market value or low grade in terms of a contained metal of high market value.

For instance, an iron ore typically ranges from over 30% low-value Fe content and sometimes above 60%, while a gold ore contains high-value gold content in ppm at say 3 g/t. Gangues are minerals present in rock either as simple inorganic or complex silicates from which no metal or non-metal of value can be extracted at present. A gangue mineral today can become an ore in future. Minerals that are commonly treated for metal extraction are called ore minerals. When different minerals bearing a particular metal occur together in a deposit, they are referred to as ore-forming minerals. Ore minerals result from certain geological processes and they are found as small, localized and isolated rock masses and such localized accumulations are

DOI: 10.1201/9781003323433-27

called mineral deposits. A mineral deposit that has sufficient concentrations of mineral values to ensure profitable extraction is called an ore deposit. The assembly or aggregate of minerals and gangues in a deposit is called an ore.

27.2   Geochemical Classification of Minerals

Based on their level of abundance in the earth's crust, metals can be grouped as geomchemically abundant and geomchemically scarce. The geomchemically abundant metals are aluminium, iron, magnesium, manganese and titanium and account for above 0.1% of the weight of the earth's crust, while the remaining metals such as zinc, lead, copper, silver and gold constitute less than 0.1% of the earth's crust. It has been observed that geomchemically abundant metals are found in rock minerals as essential minerals, while geomchemically less abundant metals are not found as essential constituents of any rock mineral. For instance, basalt comprises of olivine and pyroxene which are both magnesium iron silicate, feldspar which is calcium aluminium silicate and ilmenite which is iron titanium oxide. Although basalt contains geomchemically scarce metals, they do not constitute essential metals in any of their minerals. The geomchemically scarce metals are known to rarely form minerals in common rocks but can substitute for geochemically abundant ones.

For instance, copper, zinc and nickel can substitute for iron and magnesium in olivine and pyroxene to a limited extent. Consequently, ore minerals of geomchemically abundant metals occur in many common rocks while the occurrence of geomchemically scarce metals can only be found where some geological processes cause it. The energies required to smelt sulphides, oxides or hydroxide are less than that required for smelting silicate ores. Therefore, only a few silicate minerals can be called ores.

The factors that determine if a mineral deposit qualifies to be an ore deposit are:

a. The size of the reserves in mass
b. The grade of the mineral in terms of the metal value of interest
c. The ease of mining the mineral
d. The ease of separating the gangue minerals from the mineral value to obtain a concentrate suitable for smelting
e. The energy requirement for smelting the mineral to recover its metal content

Exploration techniques currently include exploration geology, geophysics, geochemistry and satellite imagery. In the geophysical survey, ground penetrating radar or seismic waves are used to obtain contrast in rocks to be able to select the right candidate rocks for the mineral of interest for mapping and sampling. In geochemistry, samples of surface soil are studied to determine the composition of materials underneath the earth's surface. The rise and fall of the water table cause chemical motion that makes the surface soil carry a chemical signature of the underlying materials. If the geochemical studies prove positive, then the expensive drilling stage is carried out to obtain rock fragments or cores which are sampled for testing to determine the composition of the mineral deposit.[1]

Geology refers to the study of the materials that constitute the planet earth, the natural processes acting on the materials and the resulting products from the processes. Satellite images and aerial maps of large areas of the earth can be analyzed to identify large geological structures.

27.3 Classes of Non-metallic Minerals

Non-metallic minerals are minerals that get easily broken and have no metallic luster. Non-metallic minerals are categorized as construction, refractory, industrial and gemstones non-metallic minerals. Industrial non-metallic minerals include salt, sulphur, coal, talc, phosphorite, barite, abrasives, mica, diamond, gypsum, barite, clay, feldspar, borax, phosphates, fluorite and asbestos, while refractory non-metallic minerals include dolomite, silica, magnesite and fireclay.

Furthermore, examples of construction non-metallic materials include limestone, labradorite, diorite, tuff, sandstone, marble, stone, gravel, sand and granite as well gem minerals and precious stones like emerald and sapphire. Industrial minerals are so called because they have definite industrial applications such as heat and abrasion resistance and they can be used to produce various kinds of industrial products. For instance, gypsum is used in pharmaceutical preparations, barite is used as drilling mud in the oil industry and coal is used as an energy source and to produce metallurgical coke for the blast furnace ironmaking process. In general, non-metallic minerals have the following characteristics:

1. They are not suitable for the extraction of metal. For instance, kaolin as an aluminosilicate that contains aluminium but in a complex structure and low concentration such that its Al content cannot be economically extracted. On the other hand, bauxite contains Al as simple inorganic Al_2O_3 and in a sufficiently high concentration of Al that it can be processed as an Al ore by hydrometallurgy to recover Al metal.
2. Unlike metals that can be re-melted and recycled, non-metallic minerals cannot be re-melted and recycled.
3. Non-metallic minerals are also classified as chemical, metallurgical raw materials, ceramic and glass materials, raw materials to produce binders.

27.3.1 Chemical Raw Material

Examples of chemical raw materials are natural salt, halite, borosilicates, apatite, nitrates, celestite, sulphur, sylvanite, carnalite, bischofite, polyhalite and most of them find use in fertilizer production.

27.3.2 Metallurgical Raw Materials

Metallurgical raw materials include refractory minerals, fluxes like dolomite, limestone, fluorite, quartzite and moulding materials like clays, sand and bentonite clays.

27.3.3 Construction Non-metallic Materials

Construction non-metallic materials include limestone, labradorite, diorite, tuff, sandstone, marble, stone, gravel, sand, and granite as well as gem minerals and precious stones like emerald and sapphire. Ceramic and glass raw materials such as kaolins, feldspar, wollastone, rhyolites and clays of high melting points. Raw materials for the production of binders such as clays of low melting points, limestone, marl and mineral dyes like ochers and colcothan. Thermal and acoustic insulation materials such as perlite and vermiculite.

27.3.4 Non-metallic Ore Raw Materials

These are grouped as industrial materials such as diamond, piezo quartz, muscovite, agate and phlogopite. Precious and semi-precious stones such as emerald, topaz, ruby, diamond jewellery, agate, malachite, jasper, amber and turquoise as well as asbestos, graphite, abrasives, corundum and emery are other examples.

27.4 Processing of Non-metallic Minerals

27.4.1 Processing of Phosphate Ores

Phosphate ores are used in the production of fertilizers. All phosphate ores are generally low in Mohr hardness and thus belong to the group of brittle to moderately crushable ores with good grindability. Phosphate rocks are subjected to primary and secondary crushing using jaw and cone crushers, respectively; while the screening is carried out with circular or linear vibrating screens. The crushed phosphate ores are typically subjected to one-stage closed circuit grinding in ball mills. The JXSC ball mills have been reported to reduce steel ball consumption by between 20% and 30% and are equipped with a super wear-resistant rubber liner that can elongate the ball mill service life and its rate of operation. The phosphate ore slurries are typically treated using various froth flotation techniques depending on their properties.

The direct froth flotation can be used to treat the silicate-bearing phosphorite and the metamorphic silicon-calcium apatite of sedimentary origin to remove silicate gangue. The sedimentary calcium-magnesium phosphate rocks are treated by single-stage reverse flotation to remove calcium-magnesium gangues. A combination of reverse flotation followed by direct flotation and two-stage reverse flotations can be used to obtain a concentrate enriched with phosphorous minerals by removing carbonate minerals and silicates. Phosphorite concentrate grade in terms of P_2O_5 content can reach 30–40% at a recovery of 75–90%. Figure 27.1 shows a typical phosphate ore, while Figure 27.2 present a typical flowsheet for the processing of phosphate ores.[2-4]

Figure 27.2 presents a typical flowsheet for the concentration of phosphate ore.

27.4.2 Processing of Diamond Ores

Diamond ores are found generally in kimberlite rocks which also typically host other minerals like garnets, chrome and diopsides. Diamond is found in Russia, Botswana,

FIGURE 27.1   A typical phosphate mineral—Hannayite.

Source: **Courtesy of Wikimedia (2022)**

South Africa and Canada. There are five main steps in the processing of diamond ores and these are as follows:

The kimberlite run-of-mine ore lumps are subjected to crushing in stages to gradually reduce them to sizes less than 25 mm in order to free the diamond particles from the gangue minerals in kimberlite. The crushing is applied in such a manner that the resulting products will not be too small in order not to break the diamond particles.

After each crushing stage, the wet ore is subjected to wet screening on vibrating screens with the correct screen size panel to ensure effective separation. The wet screening is preferable as it eliminates the need for dust control.

In scrubbing, the ore is washed with water to break the lump clay materials. After scrubbing, the ore is screened to remove the clay slimes and materials including diamond particles with sizes below 1 mm. Diamond particles of less than 1 mm in size are considered of no value.

The crushed, screened and scrubbed ore is treated with a Dense Medium Cyclone separator to remove the heavy minerals like diamond mineral value from the light mineral gangues like quartz. The ore is mixed into a ferrosilicon medium slurry and fed into the Dense Medium Cyclone. In the process, the heavy minerals sink and escape through the cyclone's bottom unto a sink screen, while the light minerals float and exit the cyclone into a floating screen. The diamond in the sink concentrate

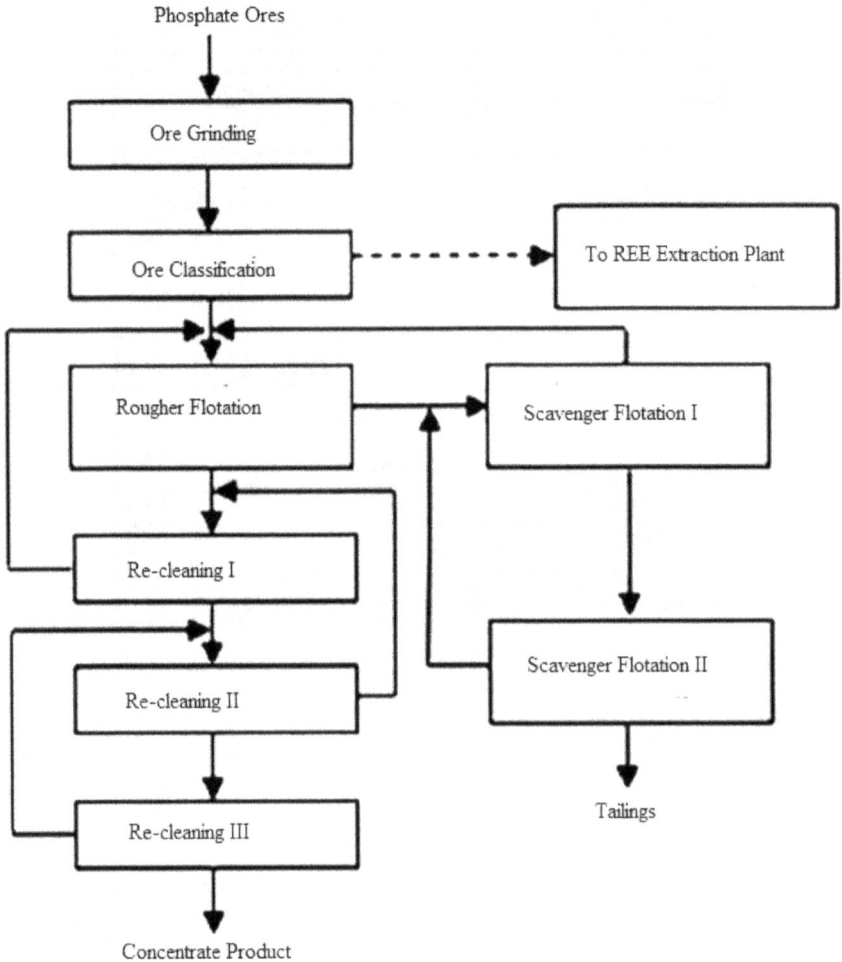

FIGURE 27.2   Flowsheet for the processing of phosphate ores (Elbendari et al., 2019).

is then recovered by other processes. The ferrosilicon medium is recovered from residual wastes for re-use by subjecting it to a magnetic concentration process.[5]

The diamonds in the sink concentrates are recovered by methods such as grease tabling described in Section 13.2.1[6] and by X-ray sorting which identifies diamond particles by their atomic mass or as they fluoresce.

Figure 27.3 shows a flowsheet to concentrate a diamond ore.[7]

27.4.3   Tar Sands

Tar sand which is also known as oil sand, is a combination of clay, sand and water and is saturated with a dense and highly viscous form of petroleum referred to technically as bitumen. Tar sands yield a mixture of liquid hydrocarbons, which apart from

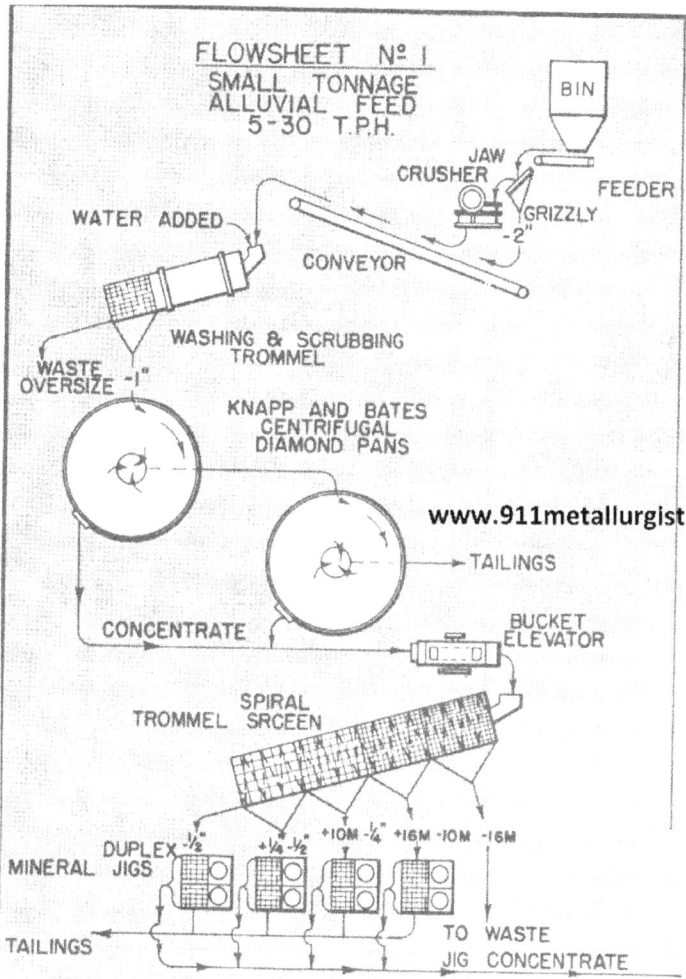

FIGURE 27.3   Flowsheet for the beneficiation of diamond ore.

Source: **Courtesy of 911 Metallurgist (2022)**

mechanical blending requires additional processing before becoming finished petro-leum products. Tar sand deposits are found in various parts of the world including Canada, Madagascar, Venezuela, Russia, the United States and Nigeria. Tar sand is exceedingly rich in oil and other valuable minerals and metals in varying proportions. Nigeria has a large deposit of natural bituminous tar sand. It is estimated that Nigeria's tar sand is the fourth largest in the world after Canada, Russia and Venezuela. The esti-mated tar sand reserves in Nigeria are about 34–45 billion barrels of heavy oil trapped in tar sand deposits in Ondo State alone, with more reserves in Edo and Ogun States.

Bituminous materials, apart from their use in the production of crude oil, are useful in the petrochemical industry. The sulphur content of tar sand constitutes a potential source of elemental sulphur for sulphuric and superphosphate fertilizer production plants. Other industrial applications include the production of anticorrosive coatings, protection for electric cable, bond adhesives and brake lining among others. Bitumen extract from tar sand is also used as an additive in the production of metallurgical grade coke. Rutile, a major source of the element titanium is also found in the residual sand of tar sand.

27.4.4   Processing of Fluorite Ores

The brittle fluorite mineral with a Moh hardness of 4 also known as fluorspar is a common mineral in nature and it consists mainly of calcium fluoride (CaF_2). It occurs in hydrothermal veins and is also commonly found in igneous rocks like granites and carbonate rocks, alkaline intrusive rocks and near the jet holes around volcano. It is commonly found in association with minerals like sbarite, spinel, albite, dolomite, chalcopyrite, scheelite, apatite, topaz, cassiterite, wolframite, pyrite, calcite, sphalerite, lapis lazuli, muscovite, quartz, rhodochrosite and galena.

Fluorite ores are found as single types and associated types. The associated types include those associated with quartz, calcite and sulphide. The single type fluorite is of high-grade with low contents of quartz and calcite but low in reserves while the associated ones are of low grade but large in reserves. The processing of fluorite ores typically involves crushing, sieving, grinding, grading, flotation, filtration and drying.

The calcite type fluorite ore typically contains more than 30% calcite with fluorite as the major constituent. Since both calcite and fluorite are calcium-based minerals, they exhibit physico-chemical surface properties that are similar and in solutions, they are subjected to the same mutual chemical transformations making it difficult to separate them. Both fluorite and calcite can be floated together with fatty acid collectors at the basic pH between 8 and 9.5 but in a weak acid medium, calcite is less floatable than fluorite. Furthermore, depressants like sodium silicate and trapping agents like oleic acid can be used to make the flotation separation between both more effective.

The barite type fluoride ore contains barite of between 10% and 40% and fluoride in association with pyrite, galena, sphalerite and other sulphide minerals. However, barite and fluoride have similar floatability. The flotation of the ore is done in two stages. Firstly, both barite and fluorite are floated combined and in the second stage, barite is separated from fluoride. For the first stage, sodium carbonate is used as a pH regulator, oleic acid as the collector and sodium silicate as the depressant. In the second stage, barite is depressed with depressants like dextrin. Afterwards, fluorite is depressed with depressants like barium chloride with oleic acid as the collector for barite. The pH regulator used in the latter stage is sodium hydrate.

For the quartz-type ore, the ore grinding level is very important. Coarse grinding will leave part of the fluorite minerals in the lumps, while over-grinding will cause part of the very fine fluoride particles to be misplaced into tailings. Rod mill grinding produces a more uniform product and the flotation yields less quartz content in the concentrates. Fluorite concentrates at over 97% grade and above 69% recovery have been obtained from quartz-type ores at a slurry pH of 9–10 and sodium silicate as the depressant.[8]

Table 27.1 presents selected non-metallic minerals occurrences and areas of applications.[6]

TABLE 27.1
Selected Non-metallic Minerals

S/N	Material	Main Mineral	Formula	Occurrence	Applications
1	Anhydrite	Anhydrite	$CaSO_4.2H_2O$	Found in saline residues with halite and gypsum	Find use in fertilizer, cement, plasters, sulphates and sulphuric acid making
2	Mica	Muscovite	$KAl_2(AlSi_3O_{10})(OH,F)_2$	Found widely in igneous rocks, e.g., granite; metamorphic rocks, e.g., gneisses; and sedimentary rocks, e.g., sandstones	Applied for insulating purposes in electrical systems
3	Fluorspar	Fluorite	CaF_2	Found as hydrothermal veins and replacement deposits	To produce hydrofluoric acid, as a flux in steel making etc.
4	Phosphates	Apatite	$Ca_5(PO_4)_3(F,Cl,OH)$	Main component of fossil bones in sedimentary rocks etc.	Mainly used in fertilizer making
5	Garnet	Pyrope	$Mg_3Al_2(SiO_4)_3$	Found in river and beach deposits etc.	Main use as abrading material in sandblasting etc.
6	Serpentine	Serpentine	$Mg_3Si_2O_5.(OH)_4$	Formed as a secondary mineral from a primary mineral such as olivine	Fibrous types are used in asbestos making etc.
7	Calcium carbonate	Calcite	$CaCO_3$	Occurs in veins with metallic ores	Used as a smelting flux, for construction etc.
8	Borates	Borax	$Na_2B_4O_7.10H_2O$	Formed as precipitates when water from salty lakes in arid areas evaporates	For use in tanning, soap, glue and cloth-making industries, fluxing in glass making etc.

(Continued)

TABLE 27.1   (Continued)
Selected Non-metallic Minerals

S/N	Material	Main Mineral	Formula	Occurrence	Applications
9	Barytes	Baryte	$BaSO_4$	Found as replacement deposit of limestone; as gangues in vein deposits of lead, copper and zinc in association with calcite, fluorite and quartz	To produce barium chemicals; as weighting agents in oil and gas drilling muds
10	China clay	Kaolinite	$Al_2Si_2O_5(OH)_4$	Formed from the altering of primary aluminosilicates especially alkali feldspars	In use as filler in paint, paper and rubber making; to produce china and porcelain wares
11	Corundum		Al_2O_3	Found in impure form as emery with high contents of magnetite and hematite; original components of igneous rocks such as syenite and metamorphic rocks such as marble; also found as alluvial deposits	Colored types serve as gemstones and abrasives because it is second only to diamond in hardness
12	Cryolite	Cryolite	Na_3AlF_6	Found in pegmatite veins in granite	Used as flux in the electrolytic extraction of Al
13	Dolomite	Dolomite	$CaMg(CO_3)_2$	A rock-forming mineral; found as gangues in veins bearing galena and sphalerite	Used as a furnace lining refractory; for fluxing in steelmaking; in building
14	Epsom salts	Epsomite	$MgSO_4.7H_2O$	Found encrusted on walls of caves and mine workings; occurs in the oxidized zones of arid pyrite deposits	For use in tanning and medicines

Source: Wills and Napier-Munn (2006) (courtesy of Elsevier Book Publisher, 2022)

References

1. Zaghlol, K. (2019): *Geological and Geochemical Exploration Techniques, Geological and Geochemical Exploration of Copper Mineralization in Egypt* (https:// www. researchgate.net/publication/336128574_GEOLOGICAL_AND_GEOCHEMICAL_ EXPLORATION_TECHNIQUES).
2. JXSC (2021): *Phosphorite Ore Mining Solution—Mineral Processing* (www.mineral dressing.com, Accessed 19th November, 2021).
3. Wikimedia (2022): *Hannayite* (https://upload.wikimedia.org/wikipedia/commons/5/55/ Hannayite.jpg, Accessed 4th May, 2022) [By David Hospital—Own work, CC BY-SA 4.0, https://commons.wikimedia.org/w/index.php?curid=75940424].
4. Elbendari, A., Potemkin, V., Aleksandrova, T., and Nikolaeva, N. (2019): Mineralogical and Technological Aspects of Phosphate Ore Processing. In: Glagolev, S. (ed.), *14th International Congress for Applied Mineralogy (ICAM2019). ICAM 2019.* Springer Proceedings in Earth and Environmental Sciences. Cham: Springer (https://doi. org/10.1007/978-3-030-22974-0_14).
5. Canada Multotec (2021): *The Five Main Steps in Processing Diamond Ore* (https:// www.multotec.com/en/distributor/canada/multotec-canada, Accessed 20th November, 2021).
6. Wills, B.A., and Napier-Munn, T.J. (2006): *Mineral Processing Technology-An Introduction to the Practical Aspects of Ore Treatment and Mineral Recovery*, 7th Edition. Amsterdam: Elsevier Science and Technology Books.
7. 911 Metallurgist (2022): *Diamond Processing Flow Chart of Beneficiation* (www. 911metallurgist.com/blog/diamond-mining-process-flow-chart, Accessed 28th May, 2022), Accessed 19th January, 2023).
8. MES (2022): *Fluorspar Beneficiation Process Plant* (www.miningmes.com/upload/ 2017-03/15/TheFluorsparBeneficiationProblemBeijingHotMiningTechCo.jpg, Accessed 28th May, 2022).

28 The Operation of a Quarry

28.1 Introduction

A rock quarry is a plant where rock boulders are crushed to smaller rock sizes and it is also known as "surface mine", "pit", "open pit" or "open cast mine". Quarries may differ based on the type of rocks mined, the size of the operation and the major modes of transport to deliver products to the customers. A quarry plant is usually located on a site where exploration geologists have identified sedimentary, igneous or metamorphic rocks which are used in buildings and structures construction. For transport cost reduction for the heavy rocks, a quarry is usually located near the community it serves. After selecting the location and getting a site design for the safe and efficient operation of the plant, operating permits that includes satisfying environmental regulations have to be obtained. The site is then cleared of its vegetation and overburdens in order to make the rocks accessible.

A mine is different from a quarry as the former involves underground mining while the latter refers to surface mining or a site for extracting minerals without a roof. Mining is associated with sites where minerals are taken out to extract metals and coals.

There are eight steps for sustainable quarrying. These are:

1. The making of a mine map to guide the activities on the site
2. The removal of the top layer materials, called overburden from the site
3. Drilling of holes and insertion of explosives and blasting of rock surface to loosen its materials
4. Transporting of materials to the plant for processing
5. Rocks are processed by crushing, screening and other equipment
6. Transport of finished products to the customers
7. Recycling or re-using of products
8. Choosing crushers suitable to produce aggregates suitable for use in construction

The main products of a quarry are called aggregates of crushed rocks, gravel and sand. Quarries are essential because they provide the materials for the construction of buildings and structures.[1]

28.2 Mining the Rocks

The first stage in mining the rock is drilling and blasting. Holes are drilled into the rock body and explosives are placed inside them to be carefully detonated to cause the minimum release of energy for the most efficient blasting. The blasting process is monitored with special equipment that records the magnitude of sounds

DOI: 10.1201/9781003323433-28

and vibrations to ensure it is within the acceptable limits in order to protect the community from noise pollution. The frequency of blasting may be daily or twice a week depending on the size of the plant. When large pieces of rocks are taken out of the earth, the area left behind constitutes the pit or the quarry. The disintegrated rocks are moved out of the pit into large haulage trucks and transported to the plant where they are processed by crushing and division into different sizes.

28.3   Crushing the Rocks

The rocks with sizes between 450 and 700 mm are delivered to the primary stone crusher such as that of the FOTE model which can take between 300 and 2000 tph with discharge in the size range of 150–300 mm. Based on the rock sizes required, the discharge from the FOTE primary crushing stage is taken into the smaller secondary crusher such as the FTM model once or more times and carried on conveyor belts for the process. The rocks can be pebbles, limestone, granite, basalt, gypsum, gravel, dolomite, phosphate and over 200 other types and the plant capacity can vary between 1 and 1000 tph. The quarry rock crushing equipment is commonly used for the primary, secondary and tertiary crushing in the plant. The trucks and trains are weighed with and without loading to determine the weight of the material to be carried out of the quarry. This will enable correct billing and ensure that overloads beyond what the law stipulates for the durability of roads and rail tracks are removed. The stationary crushing line solution can be of the following types:

Jaw crusher plus Cone crusher are used for materials with high hardness and the product discharge is uniform and of good shape, while jaw crusher combined with impact crusher is used for materials of low and moderate hardness. A hammer crusher combined with a sand-making machine can be used to produce fine-sized materials. There are also mobile quarry stone crusher units. The FTM jaw crusher combined with the impact crusher has been used to crush limestone feed sizes less than 700 mm and a discharge with sizes ranging between 150 and 400 mm at a throughput of 100 tph. The jaw, cone, impact and hammer crushers by Eriez and other manufacturers can also be used.[2]

28.4   Storage

The plant typically has large stockpiles of rocks, gravel, sand and other materials that can be up to 9.14 m in height and 243.84 m in width. Bulldozers and front-end loaders are used to keep the stockpiles in place, while customers' trucks are filled up with rocks and aggregate by a shipping loader.

28.5   Environmental Protection

The water used in the plant from rain and other sources is recycled in a closed-loop system. The water is kept in a recycling pond and its sediment is allowed to settle. The water is only released from the pond after testing to confirm it is safe and satisfies quality compatible with the environment.

28.6 Aggregates

Aggregate is the fundamental building block of any construction. A range of aggregates is produced in quarries for concrete, plumbing, asphalt spray sealing and drainage. The durability of the aggregates depends on the type of rocks from which it is produced. Aggregates can be of the following types as produced at the USA.[3, 4]

A. Aggregate of sizes 7, 10, 14 and 20 mm which are used for road construction in wet areas and as sheets over roads made with gravel. They have also been used for asphalt and drainage as well as sealing aggregate.
B. Aggregate of size range 20–40 mm which are used for road construction in wet areas and as sheets over roads made with gravel.
C. Yarck aggregate of size range 40–80 mm which is used for roadmaking in areas and tracks that are boggy as well as for drainage.
D. Ballast-Railway aggregate of size 63 mm which are used for railway and tracks that are boggy.
E. Yarck aggregate of size 150 mm which is used for gutters and gabion baskets. Secondary aggregates are those produced by crushing used construction materials made from primary materials such as concrete, bricks and asphalt. They exhibit low strength and are thus considered inferior to the primary aggregates. They are used as filler materials in constructing roads and to carry out maintenance works on laid asphalts.

References

1. Institute for Quarrying (2021): *What Is Quarrying?* (www.quarrying.org/about-quarrying/quarrying-explained).
2. FTM (2021): *Quarry Stone Crusher—Fote Machinery(FTM)* (www.foteinfo.com, Accessed 22nd November, 2021).
3. LS Quarry (2021): (www.lsquarry.com.au, Accessed 20th December, 2021).
4. The Quarry Story (2021): *What Is a Quarry? How Are They Built? How Do They Work?* (Blogs—nScale.net, Accessed 20th November, 2021).

Index

For Product Safety Concerns and Information please contact our EU
representative GPSR@taylorandfrancis.com
Taylor & Francis Verlag GmbH, Kaufingerstraße 24, 80331 München, Germany